Next-Gen Cybersecurity with Modern AI

Next-Gen

Cybersecurity with Modern AI

| LEARN GEN AI CONCEPTS
| CYBERSECURITY BEST PRACTICES WITH AI
| AI-POWERED ANTI-MALWARE SOLUTIONS
| AI ETHICS IN CYBERSECURITY
| AI TOOLS IN CYBERSECURITY

By
Dr. Milton Mattox | Ajit K Jha

SynergyBiblio

Disclaimer

Every effort has been made in the preparation of this book to ensure the accuracy of the information presented. However, the information contained in this book is sold without warranty, either express or implied. Neither the authors; nor the publisher or its dealers and distributors; will be held liable for any damages caused or alleged to have been caused directly or indirectly by this book. No warranty may be created or extended by sales or promotional materials. The advice and recipes contained herein may not be suitable for everyone.

The fact that an individual, organization, or website is referred to in this work as a citation and/or potential source of further information; does not mean that the author endorses the information, the individual, organization, or website; that it may provide or recommendations they/it may make. Further, readers should be aware that Internet websites listed in this work might have changed or disappeared between when this work was wri◎en and when it is read.

Adherence to all applicable laws and regulations, including international, federal, state, and local governing professional licensing, business practices, advertising, and all other aspects of doing business in any jurisdiction in the world is the sole responsibility of the purchaser or reader.

ISBN Print: 978-1-960911-55-1

ISBN eBook: 978-1-960911-56-8

First Edition: June 2024

Published by: Synergy Biblio Publishers
3104 E. Camelback Road #2040
Phoenix, AZ 85016
inquiry@synergybiblio.com
www.synergybiblio.com

Acknowledgment

We offer our deepest appreciation to the incredible teams at the United States Artificial Intelligence Institute and the United States Cybersecurity Institute for their unwavering support during the production of this book. Their expertise, dedication, and enthusiasm have been instrumental in shaping the content and ensuring its relevance and accuracy. Their constant encouragement and assistance made completing this project possible.

We also express our sincere thanks to our colleagues who generously reviewed drafts of the book and provided invaluable feedback. Their insights, suggestions, and constructive critiques have greatly enhanced the depth of the material.

We thank our family and friends for their steadfast and unconditional support throughout the writing process. Their encouragement, patience, and belief in this endeavor have been a constant source of inspiration. From the late-night brainstorming sessions to the countless drafts and revisions, their understanding and motivation have been invaluable.

Preface

In the rapidly evolving landscape of technology, Artificial intelligence (AI) has emerged as a transformative force, reshaping industries, and societies alike. As organizations strive to harness the potential of AI to drive innovation, efficiency, and competitiveness; the role of effective AI project management has become paramount. This book is a comprehensive guide to navigating the intricate realm of AI project management, offering insights, strategies, and best practices to successfully lead AI initiatives from conception to deployment.

In an era where technology is an integral part of our daily lives, the role of cybersecurity cannot be overstated. As we move deeper into the digital age, the threats we face are becoming highly sophisticated, requiring advanced solutions to protect our data, privacy, and infrastructure. This book, **"Next-Gen Cybersecurity with Modern AI"** is a comprehensive guide that delves into the essentials of cybersecurity while exploring the transformative role of AI in this critical field.

The rapid evolution of technology has not only brought about unprecedented opportunities but also a complex landscape of cybersecurity challenges. Traditional methods of securing networks and systems are increasingly being outpaced by the ingenuity of cybercriminals. In response, AI has emerged as a powerful tool, enhancing our ability to detect, prevent, and respond to cyberthreats with unparalleled precision and speed.

With McKinsey reporting the cybersecurity costs soar by 15% annually, they are projected to reach nearly **$10.5 trillion by 2025**. Despite this, people are not investing enough in cybersecurity.

This book is sponsored by the United States Artificial Intelligence Institute (USAII®), the world's leading AI certifications provider. The book is designed to provide readers with a solid foundation in cybersecurity principles and practices, as well as an in-depth understanding of how AI is revolutionizing these methods. From the basics of securing networks and data to the advanced applications of machine learning and deep learning in threat detection, we cover a wide array of topics that are essential for both beginners and seasoned professionals.

Next-Gen Cybersecurity with Modern AI is structured to offer a blend of theoretical knowledge and real-world insights. The chapters guide readers through the fundamental concepts of cybersecurity, followed by a preliminary perspective of AI-driven techniques and their real-world applications. Whether you are an IT professional, a cybersecurity enthusiast, or a student aiming to specialize in this dynamic field, this book equips you with the knowledge and skills to navigate and secure the digital world.

By writing this book, our goal is to provide a resource that not only informs but inspires innovation and proactive thinking in tackling cyberthreats. As you embark on this journey through the pages of **Next-Gen Cybersecurity with Modern AI**, we hope you gain a deeper appreciation of the critical role AI plays in safeguarding our digital future and feel empowered to contribute to the evolving landscape of cybersecurity.

Welcome to the future of cybersecurity, where AI is not just a tool, but a game-changer.

Dr. Milton Mattox
Mr. Ajit K Jha

Prologue

At the forefront of digital innovation, a powerful synergy between cybersecurity and artificial intelligence (AI) is emerging, as highlighted by a pivotal collaboration between leading experts in the field. This work transcends a mere aggregation of insights, merging decades of hands-on experience, collective challenges, and pioneering successes in digital security and tech innovation. The foundation laid by these experiences is further deepened and augmented through the expertise of **Dr. Milton Mattox** and **Ajit K. Jha**, each offering their distinct viewpoints and vast knowledge of the narrative.

With a venerable 35-year tenure in so◎ware development, executive leadership, and cybersecurity; Dr. Milton Ma◎ox brings a depth of knowledge that only comes from years of navigating through evolutionary tech complexities. Having served in the capacities of a CTO, COO, and CEO over the decades; his hands-on expertise in cryptography and quantum computing infuses readers with a holistic view of the cybersecurity landscape. Dr. Ma◎ox's involvement in developing operating systems and leading large organizations through the intricacies of so◎ware application development and IT illuminates the multifaceted challenges and solutions inherent in securing the digital age.

Ajit K. Jha, with 25 years of extensive experience across various leadership roles in notable educational technology and cybersecurity organizations, adds a visionary perspective to the book. His expertise as a Subject Ma◎er Expert in AI, Data Science, and Cybersecurity, alongside his sheer leadership and project management expertise in technology and consulting sectors; enriches the narrative with practical insights into the application of AI within the cybersecurity field. Mr. Jha's profound grasp of the global cybersecurity ecosystem, shaped by his involvement in product development and strategic oversight, enhances the book's examination of AI-driven security strategies and the ethical dimensions of technological advancement.

Together, these eminent authors embark on a journey through the convergence of cybersecurity and AI, providing readers with a panoramic view of the present and future of digital security. They delve into the intricacies of AI-driven threat detection, and the ethical implications of AI in cybersecurity. Explore the visionary future of AI governance and policy through the lens of seasoned experts who have led the charge in navigating the digital frontier from the front.

Their combined backgrounds, spanning the creation of essential infrastructure and spearheading international efforts in cybersecurity and AI, offer a detailed context for understanding artificial intelligent ways of transforming the cybersecurity domain. This narrative is not just about technology; it is a tale of innovation, challenge, and the relentless pursuit of securing our digital world against rapid technological advancement.

As the book unfolds, it invites readers into a world where the lines between digital defense and AI innovation blur, offering unique insights into the future of cybersecurity. It demonstrates the power of collaboration and the role of sharing knowledge to tackle evolving cyberthreats. The book stands as a beacon for those navigating the complex interplay of cybersecurity and AI, encouraging a forward-looking, and ethically grounded approach to the digital challenges of our time.

The narrative emphasizes the importance of a multidisciplinary approach to cybersecurity, integrating insights from computer science, data science, and the social sciences to forge comprehensive solutions to the challenges posed by AI and other emerging technologies. It serves as a clarion call for cybersecurity professionals to embrace AI's potential while remaining vigilant against its risks. It offers a roadmap for navigating the intricate dynamics of cybersecurity and AI.

Ultimately, this book is a testament to the power of collaboration and the crucial role of staying at the forefront of technological innovation. Through the combined expertise of its authors, the book provides a comprehensive and forward-looking exploration of the future of cybersecurity, inspiring readers to play an active role in shaping tomorrow's digital world.

Table of Contents

CHAPTER 4

CHAPTER 5

CHAPTER 7

CHAPTER 9

CHAPTER 10

CHAPTER 11

Cybersecurity in the Age of IoT and AIoT 157

CHAPTER 12

Cybersecurity Workforce in the AI Era 169

CHAPTER 13

AI Governance and Cybersecurity Policy 181

CHAPTER 1

Understanding the Convergence of Cybersecurity and AI

Cybersecurity and artificial intelligence: two distinctive terms, yet highly interconnected. Today, technology is growing rapidly and getting more advanced at an unprecedented rate. It has immensely benefitted organizations as well as individual's life. Advancement in artificial intelligence technology is the backbone to the transformation of our everyday life; making the way we live and work convenient, and easier.

However, the advancement of the same artificial intelligence technology has become a matter of concern. Leaving aside the flaws in artificial intelligence, one of the biggest challenges is the misuse of this incredibly powerful technology. It has become a great weapon for cyberattacker to carry out their crimes on a large scale. Thankfully, AI can also be used to strengthen the cyber defense system of an organization.

Therefore, we can consider artificial intelligence as a double-edged sword, offering positive as well as negative impacts in the field of cybersecurity. And this tells us about the close relationship between cybersecurity and AI.

In this chapter, we will see how the use of AI is increasing in the cybersecurity domain and explore the challenges and opportunities that come across while integrating AI into cybersecurity.

1.1 The Rise of AI in Cybersecurity

The integration of artificial intelligence technology to ease out cybersecurity operations is not new. With cyberthreats becoming highly advanced and sophisticated, the convergence between AI and cybersecurity has only evolved manifold.

Earlier, the traditional cybersecurity methods relied on firewalls and antivirus software. AI used to exist at that time as well but had limited practical applications in this field due to constrained computing power and AI algorithms. However, with the advent of advanced systems, the use of AI has become more dominant.

Let us begin by considering the digital landscape as a battlefield where on one side; organizations are continuously trying to secure their valuable digital infrastructure, sensitive data, and customer trust. On the other side, malicious cybercriminals are coming up with innovative ways to gain access to an organization's systems and data. This can either be for monetary gains, or government or political purposes.

In both scenarios, artificial intelligence has emerged as a powerful weapon and transformed the way cybersecurity professionals or cybercriminals approach cybersecurity.

1.1.1 Different ways AI can be used to protect Cyberspace

AI offers great advantages in leveraging itself for cybersecurity defense. Below is an overview of how this technology can be used to strengthen an organization's cybersecurity measures. The further chapters will discuss these concepts in detail.

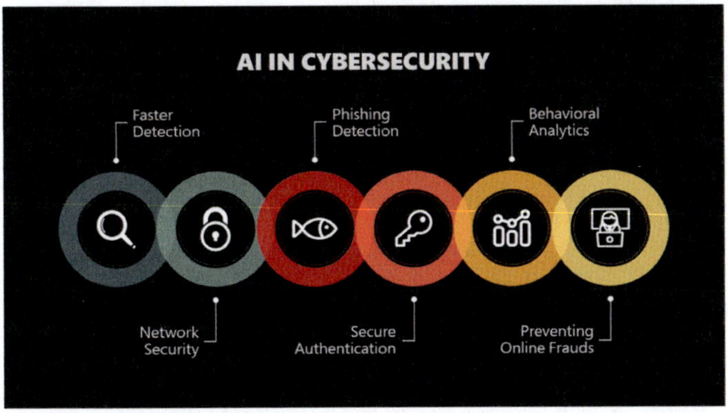

Improve Threat Detection Process

While traditional methods rely on signatures to identify threats, AI technology analyzes vast amounts of data, network traffic, user behaviors, systems logs, etc. to identify anomalies and indicate cyberattacks. This makes AI highly efficient in identifying and mitigating novel threats and zero-day attacks.

Automate Security Tasks

Cybersecurity professionals are continuously bombarded with various types of alerts out of which many turn out to be false positives. Using AI-powered cybersecurity solutions helps automate the analysis of these alerts and frees up valuable time for professionals to focus on genuine attacks. AI can automate repetitive tasks including vulnerability scanning, patch management, etc.

- **Predictive Analytics**

AI can analyze historical data and help to identify patterns resulting in predicting future cyberthreats. This helps organizations to take corrective measures on time and reduce the impact of attacks.

- **Better Incident Response**

AI helps prompt incident response by analyzing the nature and scope of security breaches in real-time.

1.1.2 Machine Learning: Powerful AI for Cybersecurity

Machine Learning (ML) is a powerful subset of artificial intelligence that can learn by itself without explicit programming. Today, ML algorithms can be trained on vast amounts of data for anomalies and threat detection, or pattern identification. The sheer benefit of using ML for cybersecurity is that it can continuously improve itself to detect and respond to threats effectively.

Here are some of the specific uses of ML in cybersecurity:

- **Anomaly Detection**

ML algorithms can analyze various parameters including network traffic patterns, user behavior, and others to identify deviations from the standard norms and flag them as potential attacks.

- **Behavioral Analysis**

This technology can analyze user behavior and identify if there are abnormal activities. So, they can easily flag compromised accounts or insider threats.

Malware Analysis

By training ML algorithms, organizations can use them to identify malicious codes by analyzing their behavior and characteristics. It is useful in identifying even newer threats.

1.1.3 The Flip Side: Challenges of AI in Cybersecurity

As artificial intelligence is being mentioned as a double-edged sword, it can be used for cyberattacks as effectively as it is used for cybersecurity defense. Cybersecurity professionals looking to leverage AI for their work must possess a strong understanding of how AI can be used for attack. This will help them plan effective defense strategies.

Currently, cybercriminals are using AI on a massive scale to carry out their attacks. They have developed newer ways to bypass AI-powered defenses and created an armed race where cyber professionals and cyberattackers are constantly innovating. Not just that, cybersecurity professionals must also keep in mind the inefficiency of AI systems as well. How well the technology performs depends on the data sets it has been trained on. Biased data can lead to AI systems exaggerating existing vulnerabilities.

Moreover, AI algorithms can be complex, and their working can be difficult to understand and know how they arrive at a particular decision. Thus, it becomes challenging to trust AI-powered security solutions due to the lack of transparency. If the AI systems are compromised, it can further worsen the situation. These could be used to launch devastating attacks. Therefore, it becomes highly important for organizations to secure AI systems properly.

1.1.4 A Glimpse at the Future of AI and Cybersecurity

Several challenges hinder the efficacy of artificial intelligence technology in cybersecurity. The rise of AI and the convergence of these domains is undeniable. Over time, when AI becomes more intelligent, advanced, and accurate; we can expect more sophisticated cybersecurity solutions capable of learning, adapting, and anticipating threats in real-time.

Organizations need to note that AI is not a silver bullet. Although largely effective, it will always need human monitoring to secure our digital world in its true essence. Human skills, knowledge, and expertise to thoroughly investigate and act smartly will always require best security practices.

Cybersecurity undoubtedly has a bright future with the integration of AI. The only thing to ensure is the responsible use of this powerful technology and continuous human intervention to keep a check on the proper functioning of the convergence of AI and cybersecurity.

1.2 Generative AI in Cybersecurity

Technology has been changing rapidly over the past few years and the advent of artificial intelligence has further accelerated this growth. The latest addition to the revolutionizing technologies is the Generative AI. Since the launch of ChatGPT and Gemini, this particular field has been taken by Storm.

The cybersecurity industry is particularly challenged every time a new technology comes in, as attackers are always in search of advanced and sophisticated ways to carry out their crimes. Though Generative AI is a great tool to contain the impact of cyberattacks, it has come up as their new weapon.

The ability of generative AI technology to facilitate new data creation makes it an incredible double-edged sword for cybersecurity. On one hand, generative AI helps security professionals proactively defend an organization's security infrastructure and data with advanced tools and enhanced threat simulations. On the other hand, it raises concerns that cybercriminals might potentially misuse it for their benefit. This chapter will give you a brief overview of Generative AI, and its positive and negative uses in cybersecurity.

1.2.1 What is Generative AI?

Generative AI, popularly known as GenAI, is the latest branch of artificial intelligence technology. It focuses on generating new content from existing data. Generative AI can generate entirely new sets of texts, videos, images, and even music, from old data. This advanced technology has proven beneficial in various tasks including data retrieval, analysis, content generation, summarization, and other growing applications.

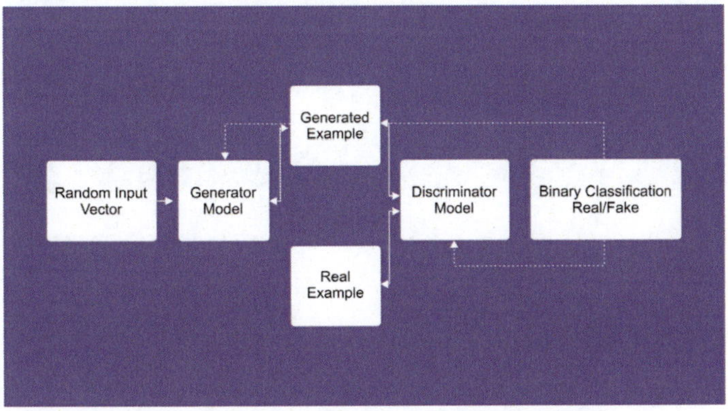

https://www.simform.com/blog/how-does-generative-ai-work/

When it comes to cybersecurity, it is a great tool that helps professionals with threat-hunting as well as providing real-time insights, critical in the vulnerability management process.

Generative AI is a part of machine learning, which is a subset of artificial intelligence. Machine Learning uses algorithms that help it learn autonomously without being explicitly programmed. It is like a sculptor who can learn a style by examining several statues. Generative AI analyzes massive amounts of datasets consisting of texts, images, codes, audio, videos, etc.

With the use of advanced algorithms to identify pa◎erns and relationships within the data. Finally, it uses the knowledge to generate entirely new content.

1.2.2 Use of GenAI in Cybersecurity for Defense

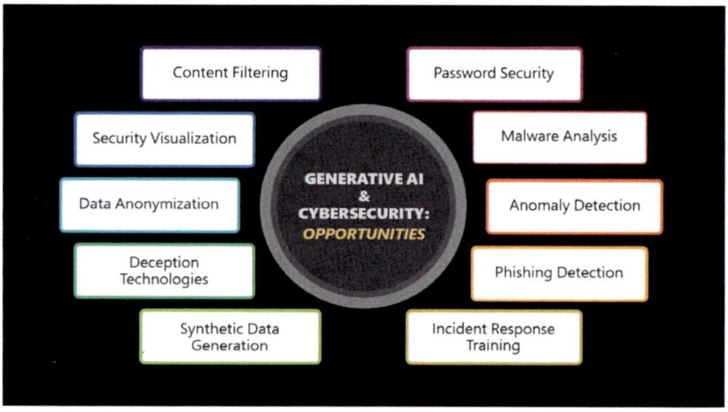

Generative AI can be used in various ways by cybersecurity professionals to strengthen their security efforts:

- **Realistic Threat Simulations**

Generative AI can create highly realistic simulations of cybera◎acks like phishing emails, malware deployment, network intrusions, etc., by analyzing vast amounts of threat data. Through these simulations, the security teams can test the efficiency of their defenses, identify vulnerabilities, as well as refine their incident response procedures in a safe and controlled environment. Thus, these simulations serve as a great way to enhance preparedness for real-world attacks.

Vulnerability management and Patch Prioritization

The volume of known software vulnerabilities is often overwhelming for cybersecurity professionals. It becomes difficult to efficiently manage and contain them. In such cases, Generative AI can help to analyze these vulnerabilities and prioritize which one to patch first. Several factors are considered like how easily the software can be exploited, existing control systems, how important the software is for business operations, and others to prioritize them.

Automated Incident response

By analyzing past security incidents and identifying underlying patterns in attack behavior, Generative AI can help automate different aspects of incident response such as containing threats, collecting evidence, etc. This automation helps to free up valuable time for professionals so that they can focus on more complex tasks like mitigation strategies or investigations.

Data Augmentation

How effective an AI-powered security tool is, rests on the quality and quantity of training data. So, Generative AI can be used to create new sets of synthetic data that resemble real-world attack patterns. Data augmentation helps to train the security models to identify even the most rare and sophisticated threats.

Stronger Passwords and Encryption Keys

Generative AI is elemental in creating uniquely complex passwords, and encryption keys that are nearly impossible to crack with brute-force methods. This is an extra layer of security that makes it difficult for cyberattackers to gain access to your data or systems.

1.2.3 Potential Negative Impact of Generative AI in Cybersecurity

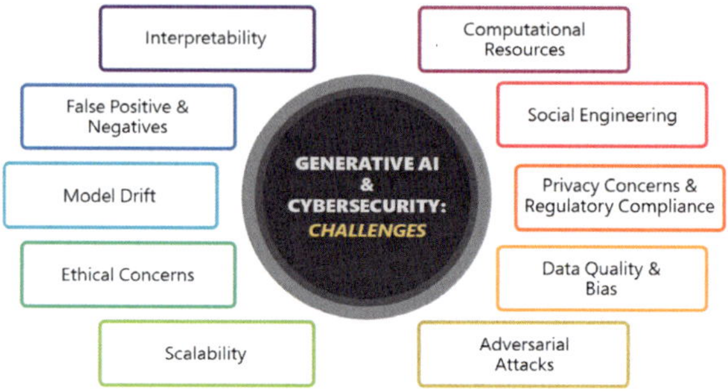

We saw how GenAI can be incredibly beneficial for security professionals. However, there are the following potential risks to consider:

- **Deepfakes and Social Engineering**

Generative AI can create realistic and convincing deepfake audio and videos. Cybercriminals can use this technology to impersonate executives or trusted individuals for social engineering a◎acks, or phishing a◎acks, trick them into revealing sensitive information, grant access to systems, or even blackmail with fake pornography.

- **Advanced Malware**

Generative AI can be used by hackers to generate newer strains of malware that can bypass traditional detection methods. The newer generation of malware can have self-evolving capabilities that can adapt its behavior and code on the fly, thereby making it highly difficult to identify and neutralize.

Bias and Discrimination

Since Generative AI models are built and trained on vast datasets, there are chances that the training datasets can contain biases and prejudices. These biases are often exaggerated in the AI models, which affect their output. It can lead to discriminatory practices in cybersecurity access controls.

Privacy Concerns

Another major concern is the privacy of users as vast amounts of data are required to train generative AI models. Therefore, it becomes important for organizations to ensure that they collect, store, and use data ethically and should follow relevant data privacy regulations.

Generative AI is a powerful tool. It can revolutionize the cybersecurity industry. Cybersecurity professionals can use it to assist them in strengthening their cybersecurity measures. However, they must pay attention to its potential misuse and privacy concerns.

CHAPTER 2

Foundations of Artificial Intelligence

2.1 AI Concepts and Terminology

Artificial intelligence at its core is a transformative technology, cybersecurity is not an exception. To be◎er understand the applications of AI in cybersecurity and how it is revolutionizing various cybersecurity operations, we must first understand the fundamental concepts and terminologies associated with artificial intelligence technology.

This chapter will help you with the core ideas underpinning AI and the essential AI terms to comprehend the AI-Cybersecurity consonance further.

2.1.1 What is Artificial Intelligence (AI)?

Artificial intelligence refers to the field of computer science that deals with creating intelligent machines that can do human tasks or mimic human cognitive functions such as learning by themselves without explicitly programming them, solving complex problems, decision-making, etc.

These intelligent machines, popularly known as AI models can easily process the given information and adapt their behavior to achieve specific goals. There are various kinds of AI ranging from basic rule-based systems to complex algorithms; capable of learning and improving themselves.

2.1.1.1 Important Characteristics of AI:

Intelligence: AI systems are highly intelligent. They can learn from data, adapt to new situations, and make data-driven decisions.

Automation: It is an important characteristic for which industries widely adopt AI systems. They can automate various repetitive and redundant tasks, freeing up human resources. Automation assists in processing information faster and more efficiently.

Learning: AI exhibits three broad kinds of learning- supervised (trained with labeled data), unsupervised (working with unlabeled data), and reinforcement learning (learning through trial and error with feedback)

2.1.2 Core Concepts in AI

Here are some core concepts necessary to understand AI functions.

Machine Learning (ML): It is a subfield of artificial intelligence that helps AI systems to learn from data autonomously without being explicitly programmed. With the help of ML algorithms, the machines can make predictions, and improve their performance over time, when they are exposed to vast data.

Deep Learning (DL): It is a type of machine learning that has been inspired by the structure and function of the human brain. They use neural networks i.e. interconnected layers of algorithms that process information in a similar manner to which human biological neurons function. They are great at performing tasks such as image recognition and natural language processing.

Data: It can be considered as the fuel of AI systems. The performance and effectiveness of AI models largely depend upon the quality and quantity of data. The better the quality of data, the better the outcome. In terms of cybersecurity, the data can be network logs, user behavior, threat intelligence, system information, etc. which are used to identify and respond to security incidents.

Algorithms: Algorithms are a set of instructions that define the steps to be followed by computers to resolve diverse problems. Algorithms are designed to process data, extract knowledge, and make decisions.

2.1.3 Types of AI Systems

Although; artificial intelligence is thought to be one whole unit, it is categorized in different ways based on its level of autonomy and capabilities.

Artificial Narrow Intelligence (ANI): ANI, also known as weak AI; is an AI system that can perform specific tasks very well but lacks general intelligence. Most of the applications in cybersecurity fall under this category and are great for tasks such as threat detection and malware analysis.

Artificial General Intelligence (AGI): AGI can be compared to intelligent machines possessing human-like intelligence or even surpassing it. They can apply their knowledge to different intellectual tasks. Though they do not exist, there is a great probability of their existence in the future.

Artificial Super Intelligence (ASI): ASI is another hypothetical AI that is believed to have higher intelligence than humans in all aspects. This concept raises significant ethical concerns for its development and deployment.

2.1.4 Benefits of AI in Cybersecurity

https://www.linkedin.com/pulse/present-future-artificial-intelligence-cybersecurity-cybalt/

Before we delve deep into the applications, workings, and uses of AI in cybersecurity, let us have a glance at the benefits AI offers in the field of cybersecurity.

Enhanced Threat Detection: AI can analyze huge amounts of data that can identify anomalies, and escape human analysts, leading to faster and more accurate security threats detection.

Improved Automation: AI can efficiently automate repetitive tasks such as log analysis, incident response procedures, vulnerability scanning, etc. This frees up cybersecurity professional's time to focus on more productive and strategic work.

Continuous Monitoring: AI systems can monitor networks and user activity continuously. This helps with constant vigilance against potential threats that often occur outside business hours.

Predictive Analytics: AI can efficiently analyze historical data and identify pa◎erns to predict future cybera◎acks. This is particularly helpful in taking proactive measures before an a◎ack occurs.

Improved decision-making: Organizations can use AI systems to analyze vast amounts of data and get insights related to security. This will enable them to make informed decisions on how to mitigate threats, allocate resources properly, and more.

These fundamental concepts of AI are indispensable in understanding the applications of AI in cybersecurity. Through this, students and professionals can build a strong foundation to explore limitless possibilities and challenges of AI in cybersecurity.

2.2 Machine Learning Algorithms in Cybersecurity

In the previous chapter, we understood the basic concepts of artificial intelligence necessary to understand its working and applications in cybersecurity. Machine learning is among the most critical AI concepts that are widely used for cybersecurity. It is the backbone of many AI-powered cybersecurity solutions.

Machine learning algorithms help the systems to learn from data, identify pa◎erns, and make predictions without explicit programming. In this chapter, we will see various ML algorithms that are commonly used in cybersecurity applications and explore how they contribute to improved security posture.

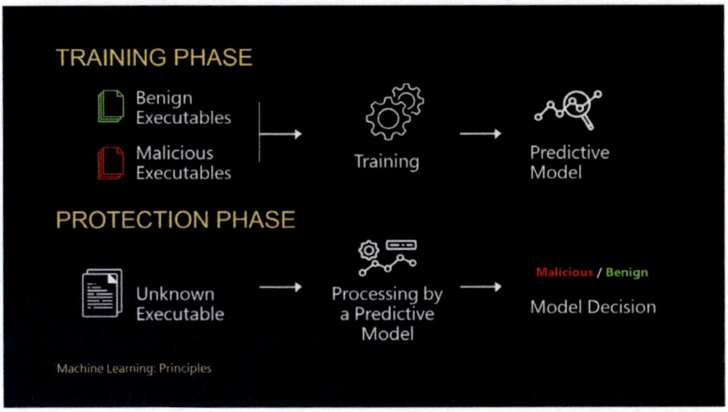

2.2.1 Supervised Learning for Threat Detection

Supervised learning algorithms are trained on labeled data sets where each data point has a corresponding label indicating the desired outcome. In cybersecurity, these labels represent malicious or suspicious activities. Some of the prominent supervised learning algorithms used in the cybersecurity domain for threat detection are:

1. **Classification algorithms:** This type of algorithm classifies data points into predefined categories. Following are the common examples of classification algorithms:
   - ***Support Vector Machines (SVMs):*** they are very effective for high-dimensional data. They create a hyperplane separating data points based on different classes. In the cybersecurity domain, it is used for separating normal network traffic pa◎erns from malicious activities.
   - ***Decision Trees:*** This type of algorithm creates tree-like structures where each note is a representation of a decision based on a specific feature of data. This algorithm decision-making process starts from the tree based on the data point's features and reaches the leaf node that represents the predicted class (for example: malicious in the context of cybersecurity).

  - *Random Forests:* This learning method combines multiple decision trees. Thus, it improves the overall accuracy and robustness as compared to the single decision tree.

2. **Regression Algorithms:** The regression algorithm is not commonly used for threat detection but it is very useful for predicting the severity or impact of a security threat based on historical data.

2.2.2 Unsupervised Learning for Anomaly Detection

Unsupervised learning is mostly related to unlabeled data sets, where the data points lack predefined categories. The goal of using unsupervised learning is to identify hidden patterns and structures within the data. In cybersecurity, it helps detect anomalies that can reveal potential threats.

Some common unsupervised machine learning algorithms are:

- **Clustering algorithms:** This algorithm groups data points into clusters based on certain similarities. In cybersecurity, clustering is used to identify groups of network activities deviating from normal patterns and indicating potential malicious activity. K-means clustering and hierarchical clustering are common examples of clustering algorithms.
- **Outlier Detection Algorithms:** This algorithm identifies data points that fall outside of the expected range. Suspicious traffic patterns, abnormal user behaviors, and other anomalies constitute outliers in cybersecurity.

2.2.3 Reinforcement Learning for Security Operations

Reinforcement learning refers to the training of AI agents/ systems through trial-and-error methods in a simulated environment. The AI systems are rewarded when they provide desired actions and get penalties for undesirable results. Over time, this helps the systems to learn and optimize their behavior to maximize the rewards. Though it is still evolving in the field of cybersecurity, reinforcement learning is used in a variety of applications such as:

- **Adaptive security systems:** Reinforcement learning agents can learn continuously and adapt to security configurations based on real-time network activity and threat intelligence. This helps to optimize the overall defense strategies of the organization.
- **Autonomous threat hunting:** AI systems can automatically explore networks on their own and identify vulnerabilities. They can autonomously take actions to mitigate them as well mimicking the role of human security analysts in an automated way.

2.2.4 Challenges of Machine Learning in Cybersecurity

Machine learning in Cybersecurity offers a ton of benefits, however, it comes with some challenges that need to be acknowledged for the proper use of AI in cybersecurity:

Data Quality and Bias

The quality and quantity of data determine how effective the ML models are. If the data is biased or compromised, it can lead to biased algorithms which might not be effective in detecting or responding to certain types of threats and may even disproportionately flag genuine activities as malicious.

Adversarial Attacks

Cybera◎ackers try to manipulate data and exploit vulnerabilities in the ML models to bypass security controls. Therefore, adversarial machine learning techniques need continuous vigilance to remain effective in the long run.

Explainability and Transparency

How machine learning models and algorithms come to a decision/conclusion is critical to understand. These are o◎en complex and lack proper explainability; which makes it difficult to trust the model's output.

2.2.5 The Future of Machine Learning in Cybersecurity

Both machine learning and the cybersecurity domain are evolving continuously. Some promising areas of ML in cybersecurity include:

- **Explainable AI**: More focus will be given to make AI models interpretable and understand how the models have arrived at a specific decision. This is important in building trust among stakeholders and ensuring the responsible use of AI for a variety of security applications.
- **Federated Learning**: This means training ML models on a variety of data sets without compromising data privacy. This will help organizations who are hesitant to share sensitive security data in a central location.
- **Updating regularly**: As the technology grows, ML models need to continuously learn and adapt to stay ahead of evolving cyberthreats as well.

Machine learning is an incredible technology and has huge potential to improve the overall security efforts by automating them and enhancing their accuracy. The further chapters will delve deeper into the applications of machine learning in cybersecurity.

2.3 Real-world Applications of AI in Different Industries

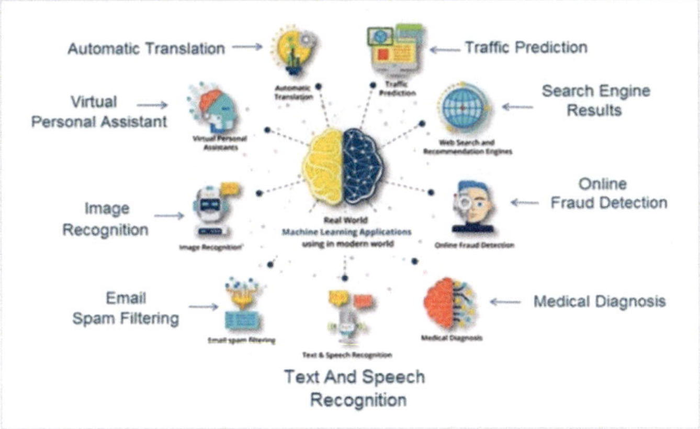

https://www.aiiottalk.com/machine-learning-examples-in-real-life/

Not just cybersecurity, artificial intelligence has been deeply embedded in our day-to-day tasks. Organizations across industries are using AI to enhance their business operations and customer experience. In the previous chapter, we saw the role of AI in cybersecurity. In this chapter let us explore how AI is revolutionizing other sectors.

2.3.1 Finance

One of the most important applications of AI in finance is fraud detection. AI algorithms can easily analyze vast amounts of transaction data and identify pa◎erns that indicate fraudulent or suspicious activities. Banks and other financial institutions can detect and prevent fraud in real-time. It helps protect both businesses as well as consumers.

Another important aspect of finance is trading. AI-powered trading platforms can use complex algorithms to analyze market trends, identify the right trading opportunities, and even support executing trading at higher speed.

Several financial institutions have started using AI-powered chatbots and virtual assistants that offer personalized financial advice based on customers' financial goals, risk tolerance, and previous investment data. Now, financial guidance is not limited by just human advisors but highly efficient virtual assistants too.

2.3.2 Healthcare

AI is proving to be highly effective in medical diagnosis and treatment. Healthcare professionals are using AI for the analysis of medical images such as X-rays and MRIs. It helps doctors diagnose diseases accurately. AI easily analyzes patient data and helps to identify potential health risks, if any. This is particularly helpful in customizing treatment plans according to patient's individual needs.

Analysis of datasets on molecular structures and biological interactions can be sped up with AI; which can lead to the discovery and development of new drugs. This helps medical researchers discover new drugs and streamline their drug development process.

AI has also been a great technology for medical surgeries. Several advanced AI-powered surgical robots are highly precise and have great control even for complex procedures. This makes faster recovery possible and improves the patient's overall surgical experience.

2.3.3 Retail and E-commerce

Artificial intelligence technology is transforming the retail and e-commerce industry with tons of features. Leading e-commerce platforms are leveraging AI to analyze their customers' browsing and purchase history, demographics, and browsing behavior to recommend personalized products. This has resulted in improved shopping experience, more customer engagement, and higher conversions.

By analyzing historical sales data and market trends, AI algorithms have proven to be highly beneficial in predicting the future demand related to a product. With this information, retailers can maintain their inventory levels and eliminate the risk of running out of stock at the time of demand.

AI is highly beneficial in analyzing real-time market conditions, competitor pricing, and customer demand so that it can come up with the right knowledge to adjust the product price dynamically. This, in turn, has helped retailers maximize profits and remain competitive in the market.

2.3.4 Transportation

Autonomous vehicles are the best example of what AI is capable of in the transportation industry. AI is transforming this section of the industry rapidly. Self-driving cars use various kinds of AI technologies including advanced computer vision and sensor fusion to navigate through roads safely by themselves. However, there are some technological and regulatory concerns, autonomous vehicles are certainly the future of the transportation industry.

AI can analyze real-time traffic data that can help identify congestion and recommend the best routes to drivers. This will help reduce traffic jams in urban areas, as well as optimize the best routes for fleet vehicles to save cost on transportation.

Another great application is the predictive maintenance of vehicles. Vehicles are equipped with a variety of sensors and by analyzing this sensor data, AI can help predict the potential maintenance issues. This will help the vehicle owners to take preventive measures on time and maximize vehicle performance.

2.3.5 Manufacturing

Just like reading and analyzing vehicle sensors, AI can analyze sensor data from machines and predict potential future failures. So, the supervisors can take preventive measures and minimize downtime along with reducing maintenance costs.

Quality control automation is another aspect of AI in manufacturing. AI-powered systems can visually inspect product defects with higher accuracy and speed. Eventually, the product quality and consistency are enhanced.

Industrial robots use AI systems that can perform complex tasks with greater precision and adaptability. They can do the heavy lifting and redundant jobs round the clock, without fatigue and quality compromise.

2.3.6 Applications in Other Industries

These were just some of the many uses of AI in various sectors. Several other industries are actively adopting AI for better services such as:

- Media and entertainment industries- AI can recommend shows based on user preferences, generate personalized content, and help with movies and game development
- In the education sector- AI can offer personalized learning experiences to students and automate students' grading process.
- AI-powered chatbots are excellent customer service executives, which can handle basic and repetitive queries and free up human executives to handle more complex queries.

These are a few examples of how AI is transforming various industries. In the following chapters, we will see the use of AI in cybersecurity in greater depth.

CHAPTER 3

Cybersecurity Basics

3.1 Key Cybersecurity Concepts

Today we live in a world where not just humans, and countries but devices are also highly interconnected. And this interconnected digital world thrives on connectivity and information sharing. However, this interconnectivity offers great convenience at work and personal levels, it also creates vulnerabilities that can be easily exploited by cybercriminals.

Not just professionals working in the field of cybersecurity, but every individual must be aware of the basic cybersecurity concepts to ensure the utmost safety and confidence while spending time and interacting on the web.

These basic concepts will help them build a robust online defense strategy and protect their data, devices, privacy, and beyond.

3.1.1 CIA Triad: Confidentiality, Integrity, and Availability

CIA Triad is the founding principle of cybersecurity. Here is what it infers:

1. Confidentiality: This principle ensures that only authorized users can access confidential information. Now, users have hundreds of accounts online; be it financial, social, informational, etc. Confidentiality means only legitimate users have the right keys and passwords to access the content and account.
2. Integrity: Data Integrity refers to complete and accurate data, which ensures there is no tampering done with the requisite data. Incorrect and forged information can lead to huge adverse impacts; both personally and in businesses. Thereby, integrity safeguards the document, and information from unauthorized modifications and helps maintain its genuity.
3. Availability: This principle aims to ensure information and systems are always available for authorized users, whenever needed. There might occur situations when information becomes inaccessible due to some or other technical issues. Availability eliminates this and ensures data is always accessible.

All security measures are aimed at achieving these three core principles of cybersecurity.

3.1.2 Authentication and Authorization

Authentication and authorization are important aspects of cybersecurity that work together to control access to systems and information.

Authentication: It is the process of verifying the user to be genuine as per the set identity. It infers the user proves himself to be the person for whom the content is intended. Some common authentication methods include passwords, PINs, fingerprint scans, facial recognition, etc.

Authorization: This is the next step after authentication. Once the user has been identified as genuine, he is granted authorization to access the content intended. It is the process of permitting users to access systems, data, and infrastructure to view, edit, modify, or even delete content/data.

Now, advanced technology like multifactor authentications (MFA), has helped to strengthen this process by requesting additional verification elements beyond just passwords, and PINs.

3.1.3 Encryption and Decryption

Securing information while its transmission from one source to another is very important to ensure it is not tampered with or is not assessed for misuse. Encryption scrambles the data using mathematical algorithms and secret keys. This scrambled data is called ciphertext and is unreadable by anyone who does not have access to the decryption key. It is like locking the message via a code and only the intended person can read it after unlocking the code.

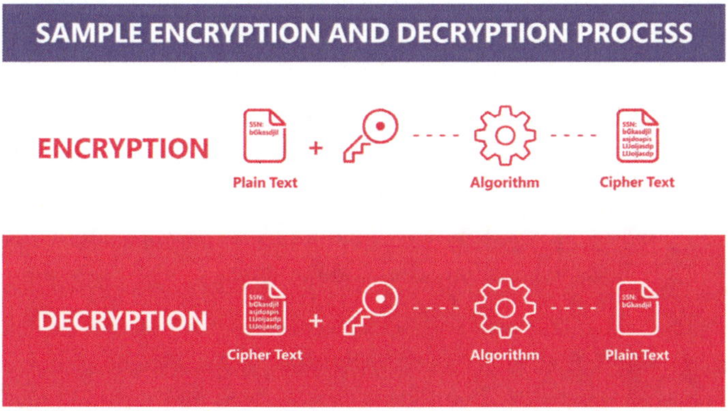

Encryption ensures data is safe during storage and transmission. This is particularly essential in safeguarding information such as financial records, personal details, confidential business documents, etc.

Decryption is needed at the end-point where the intended person needs to unlock the information. So, this process uses the corresponding decryption key and translates the ciphertext back to its original readable form.

Advanced Encryption Standard (AES), RSA, and TLS/SSL are some of the common and widely used encryption techniques for securing online connections.

3.1.4 Vulnerability Management

Vulnerabilities refer to the weak points in systems, devices, so◎ware, or networks that can be easily exploited by cybercriminals. No ma◎er how well these elements are designed, there are always ways to compromise their security. Therefore, vulnerability management becomes an important component of cybersecurity. Vulnerability management includes 3 basic procedures:

1. Identifying vulnerabilities: This constitutes regular scanning of systems and so◎ware for known vulnerabilities using various types of vulnerability scanner tools
2. Prioritizing vulnerabilities: This step assesses the severity of the vulnerabilities identified; based on the potential impact they can cause; if exploited.
3. Patching: It means installing updated so◎ware and security patches; provided by vendors and employing other methods to eliminate the vulnerabilities.

Every organization and cybersecurity professional must understand that effective vulnerability management is not a one-time process but a lifelong journey. As the technological landscape is highly dynamic, they can discover new vulnerabilities quite o◎en and therefore, prompt patching and mitigation strategies are required to reduce the a◎ack surface.

3.1.5 Network Security

Network security refers to the processes, tools, and strategies implemented to protect a computer network from unauthorized access. It involves the elimination of intrusion attempts and the insertion of malicious codes within the organization's network systems.

Some of the common ways to strengthen network security are employing strong firewall techniques, and intrusion detection and prevention systems.

Firewalls are the barriers between an organization's network and the internet. They filter the incoming and outgoing traffic based on predefined security rules and standards. Highly advanced firewall methods called Next-Generation Firewall Systems can identify and block malicious traffic attempting to enter an organization's network.

On the other hand, intrusion detection and prevention systems can continuously monitor network traffic to identify any suspicious activity that can lead to a potential cyberattack. IDS identifies and informs threats to the concerned cybersecurity professionals, and IPS activates automated prevention tasks, to eliminate attacks and alert the security team for a prompt response.

Large organizations employ network segmentation techniques that divide the network into smaller segments helping to limit potential damage in case a breach occurs.

These are some of the common network security methods. With the rise in artificial intelligence technology, these techniques have improved and become advanced and efficient in protecting an organization's networks.

3.1.6 Incident Response

No ma◎er how strong an organization's security system is, cybercriminals are innovative enough to find a way or two to breach the security. So, even if all security measures are in place, there are chances that cybera◎acks can occur at any time.

This is why companies rely on building a strong incident response plan that outlines essential steps to be taken to contain and eliminate the a◎acks. An effective incident response plan ensures a coordinated and efficient response mechanism that reduces damage and operational downtime.

Here are the key components of an effective incident response plan:

- Preparation: Developing a plan outlining roles, responsibilities, and communication protocols
- Identification and Detection: Prompt identification and detection of security breaches
- Containment: Blocking unauthorized access points, isolating devices, suspending compromised accounts, etc.
- Eradication: Immediately removing threats including malware, and patching vulnerabilities.
- Recovery: Restoring normal operations and recovering lost data through backups
- Reporting and Forensics: It involves analyzing security breaches and preparing for the future

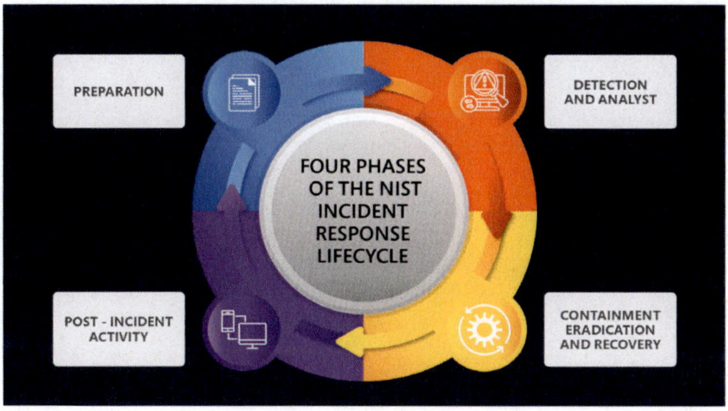

These are the basic cybersecurity concepts that help in planning, identifying, and executing security strategies efficiently. Whether you are a cybersecurity professional, organization, or individual; possessing a fundamental understanding of these fundamental concepts can help secure your systems and devices, and minimize the damage due to cyberattacks.

This basic knowledge serves as the foundation to build advanced skills in cybersecurity.

3.2 Different Types of Cyber Crimes

An email informing you won a lottery, unusual traffic on your website, and inaccessible systems, are all forms of cyberattacks.

While the traditional cybercrimes were limited to phishing attacks, and injecting malware in systems and networks. The modern cybersecurity threat landscape has evolved rapidly. Malicious actors try sophisticated and innovative ways to steal data, disrupt operations, and extort money.

So, when you plan effective security measures for your organization, you must understand different types of cybercrimes.

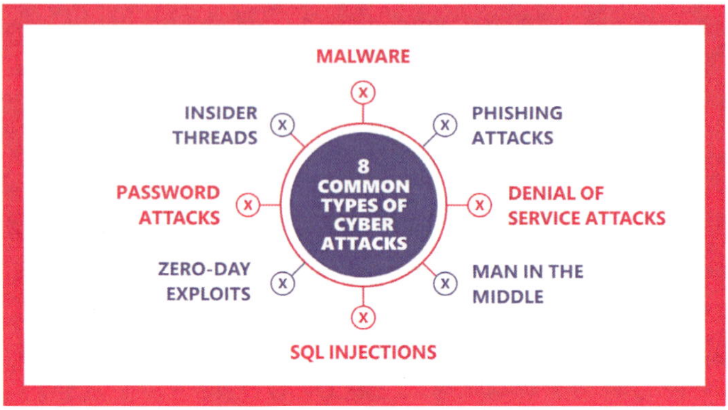

3.2.1 Data Breach

Data breach refers to the type of cybercrime in which sensitive information such as personal details, financial records, intellectual property, and other confidential data are accessed by unauthorized individuals or groups. Hackers exploit vulnerabilities within an organization's systems and networks, steal login credentials, or do social engineering a◎acks to gain access.

Impact Of Data Breaches

Data breaches can have severe consequences, not just for organizations but for individuals as well. Malicious actors can use this stolen information for various purposes including identity the◎, or financial fraud, which can severely impact individuals. Corporate data breaches can damage an organization's brand reputation and may lead to he◎y fines.

3.2.2 Ransomware Attacks

Ransomware is malicious software that can encrypt or lock a victim's data and devices, making them inaccessible. The data is then decrypted, and the system is unlocked; only after the victim pays the demanded ransom to the hackers. Payment is typically asked in the form of cryptocurrency, which makes tracing cybercriminals difficult.

Impact of Ransomware Attacks

Ransomware attacks have the potential to cripple businesses, hospitals, and even government agencies. When under a ransomware attack, the organizations are non-operational, and an additional demand for high ransom payments can result in significant financial losses.

In 2017, the WannaCry ransomware attack that affected over 200,000 computers in over 150 countries is known to be a large-scale ransomware attack. The same year witnessed the NotPetya attack, causing billions of dollars of damage worldwide.

3.2.3 DoS and DDoS Attacks

Denial of Service (DoS) and Distributed Denial of Service (DDoS) attacks are a type of cybercrime, where attackers use bots to overwhelm a website or server with massive amounts of traffic. This leads to an overload in the network infrastructure, making it inaccessible for genuine users to access the information or service.

While DoS originates from a single source, DDoS attacks involve several compromised devices to carry out the DDoS attack on a large scale and thus are difficult to mitigate.

Impact of DoS and DDoS

These types of a◎acks disrupt online services which leads to a loss in revenue along with reputational damage. Online businesses, financial institutions, and high-value service websites are frequent targets for such types of a◎acks.

3.2.4 Social Engineering and Phishing Attacks

These are the types of a◎acks that target humans by manipulating their emotions and psychology to trick them into revealing sensitive information, clicking malicious links, downloading malware, sending explicit images or videos, etc. Cybercriminals o◎en impersonate themselves as legitimate entities such as banks, social media platforms, friends, colleagues, etc. to carry out these a◎acks.

Phishing emails and pretext calls are where the a◎acker pretends to be a trusted source, and bait such as offering something desirable to lure a victim; are some of the common social engineering tactics.

Impact of Social Engineering Attacks

Social engineering a◎acks can serve as the starting point for data breaches, and ultimately lead to financial losses and identity the◎. These types of a◎acks are used to gain access to corporate networks and launch highly sophisticated a◎acks.

3.2.5 Zero-Day Attacks

Zero-day a◎acks target previously unknown vulnerabilities in so◎ware and systems. There are o◎en vulnerabilities that cybersecurity misses out on to identify and patch. These vulnerabilities pose significant threats to organizations.

Since the vulnerabilities are not known and not patched, the chances of zero-day attacks are higher. Such attacks can target individuals, businesses, and even government organizations.

3.2.6 Supply Chain Attacks

Supply chain attacks target third-party vendors and software used by organizations to gain access to the main target's network. It is like infecting the ingredients in a food store to eventually affect the recipes in the restaurants.

They prove to be highly damaging for organizations as they breach their security systems through trusted sources. These attacks can lead to data breaches, malware infections, and disruption of important business operations.

The SolarWinds Attack of 2020 is a widely recognized supply chain attack that compromised network management software and impacted thousands of organizations around the world. Kaseya's supply chain attack of 2021 is another example of a large-scale supply chain attack that infected many downstream businesses.

3.2.7 Internet of Things (IoT)

The Internet of Things is now a soaring trend and refers to the growing network of interconnected devices. Smart wearable health devices, smartphones, home appliances, and industrial control systems; constitute advanced IoT devices. These are often the weak points or vulnerabilities that hackers exploit to gain access to networks and devices.

Hackers can exploit these IoT devices to launch large-scale DDoS attacks to disrupt an organization's infrastructure or steal sensitive information. Not just the infiltration of smart home devices; it can lead to attackers spying on residents and manipulating their connected systems.

These are some of the most common types of cybercrimes. As an organization or cybersecurity professional, you must be aware of these to design and implement your organization's online security posture and protect yourself from evolving threats as an individual. Cybersecurity is an evolving domain. Therefore, it must be noted that these are not the only forms of cybercrime, as newer forms of cyberthreats shall evolve ahead.

3.3 Malware and its Types

Malware, primarily called malicious so◎ware, includes a wide range of programs designed to harm the computer and network systems or servers.

Malware comes in different types and serves diverse purposes such as stealing, encrypting, or even deleting sensitive data. They are even capable of hijacking computer systems and monitoring the end user's computer activity.

What impact malware can make at the user's endpoint depends on the type of malware and its desired goal. In some cases, its effects can be relatively mild, and in others, it can cause serious damage. Irrespectively, they are meant to exploit user's devices and networks.

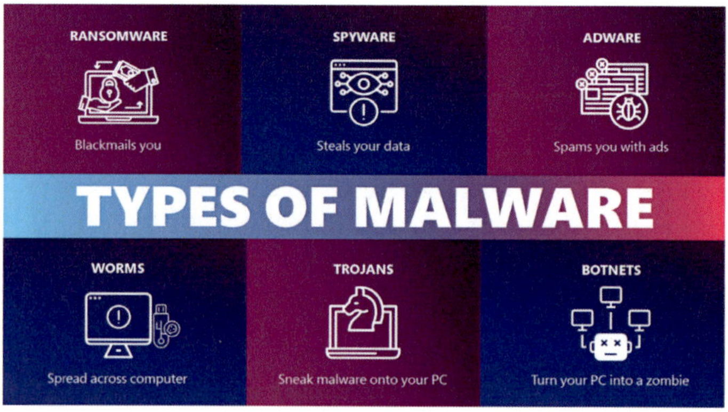

3.3.1 Different Types of Malware

Some of the most used types of malware are:

1. Adware

Adware is a type of malware used to bombard users with many unwanted advertisements, pop-ups, and banners. However, they are not destructive; but can be intrusive and heavily impact a website or device's browsing experience.

Adware can significantly slow down your computer, consume high system resources, and frequently redirect you to malicious websites.

They are often bundled with free software that you download over untrusted websites or install through deceptive tactics.

2. Spyware

As the name suggests, these types of malware spy on your devices and secretly collect information about your online activities, browsing habits, keystrokes, personal data, and more. The information collected is then either sold on the dark web or used for malicious purposes like identity theft and financial fraud.

They can be dangerous as they are designed to compromise your privacy and expose your sensitive personal information. These types of malware are often injected via malicious email attachments, compromised websites, or bundled with free software.

3. Browser Hijacking Software

Browser hijacking software can hijack your browser by changing its settings and redirecting you to unwanted websites or manipulating search results. They can significantly disrupt your overall browsing experience. They can expose you to fall victim to phishing scams, or continuously redirect you to install more malware in your systems.

4. Viruses

Computer Viruses are like biological viruses that can self-replicate and attach themselves to different computer files and applications. Once executed, computer viruses replicate and spread to other files and computers in the network to cause huge damage by impacting their normal functioning.

Viruses can corrupt or delete system files and make the computer systems unusable. Some of them can even steal confidential information and download other malware. They are spread like spyware through email attachments or infected websites, and removable media devices such as USB pen drives, and other techniques.

5. Worms

This malware is like viruses which can multiply. They do not necessarily require attaching to existing files.

Worms can exploit network vulnerabilities and spread from one computer to another. They are designed to consume computer resources and create disruption within networks. Their self-replicating capability and movement across networks without external assistance make them more disastrous.

6. Trojan Horses

The Trojan horse got its name from the epic Greek Mythology, where a wooden horse named Trojan was used to enter the city of Troy. The computer trojan horse disguises itself as legitimate software or files and tricks users to install them. Once installed, they can perform different types of malicious activities to affect systems and networks. They can steal system data, download other malware, or create backdoors for remote access.

They can steal your system's sensitive information, financial information, login credentials, etc. leading to severe losses and identity theft.

7. Scareware

As per their names, scareware uses user's fear and deception to manipulate them into downloading unwanted software or parting with money. Scareware bombards users with fake security alerts and claims to have detected critical security threats on their devices.

They trick users into downloading malicious software disguised as legitimate ones and may lead to financial losses as the users pay for fake customer support scams.

3.3.2 Detection of Malware

There are various ways that users can identify if their systems and devices are infected with malware.

Pay attention to:

- Slow system performance: The infected computer may take time to load programs and browse the internet. This is because malware may be consuming system resources.
- Suspicious pop-up ads and alerts: Malware leads to frequent pop ads and unnecessary security warnings asking users to download software.
- Unknown Programs and browser changes: You may notice your browser has changed automatically and find unusual programs automatically installed in your system if infected by malware.

- Deleted and corrupt files in your system: If you find important files missing or ge◎ing corrupted, it is a sign of malware activity.
- Frequent system crashes and network issues: Malware communicating with remote servers may lead to frequent network crashes and unexpected spikes in data usage.

If users face any of these signs, it is recommended that they immediately scan their system with a trusted antivirus program to detect and remove potential threats.

3.3.3 Removal of Malware

Commonly, the systems get infected by different types of malware. Therefore, there are diverse ways to remove them as well.

1. Scan the system with a reputable antivirus program that can identify and remove malware if found.
2. System scanning in safe mode can help delete malware that cannot be deleted in normal mode.
3. Advanced users can use specific tools and instructions from antivirus vendors to remove complex malware programs.
4. It is highly recommended to seek assistance from professionals who can restore normal functioning.

Though systems can be easily restored a◎er being infected by malware, they can cause damage to important files. Therefore, prevention is always be◎er than cure and users must try to avoid ge◎ing their systems infected by such malware.

3.3.4 Prevent Malware Infections

Preventive measures can be taken beforehand to prevent systems and devices from ge◎ing infected with malware.

- Keep the system so◎ware including operating systems, applications, and firmware up-to-date with the latest security patches
- Only download from trusted sources and avoid clicking suspicious links. Do not open links and a◎achments from unknown senders as they are the major source of malicious links
- Always use antivirus so◎ware in your system from reputable vendors to detect and eliminate malware in real-time
- Beware of social engineering a◎acks including unsolicited emails, calls, and messages; that might pressure you to click on the links, download a◎achments, or ask you to reveal personal information.

These basic protective measures can help your systems stay uninfected and safeguard confidential information.

Key Cybersecurity Concepts

- The CIA Triad
- Authentication and Authorization
- Encryption and Decryption
- Vulnerability Management
- Network Security
- Incident Response

Different Types of Cybersecurity Crimes

- Data Breaches
- Ransomware
- DoS and DDoS
- Social Engineering Attacks
- Zero-Day Attacks
- Phishing and Social Engineering
- Supply Chain Attacks
- Internet of Things (IoT) Threats

Malware and its Types

- Adware
- Spyware
- Browser hijacking software
- Virus
- Worms
- Trojan horses
- scareware

Cybersecurity Best Practices

- System Security
- Software Security
- Infrastructure security
- Hardware security

CHAPTER 4

AI-Driven Threat Detection and Prevention

Artificial intelligence has become an indispensable strategy to identify and mitigate the advanced forms of cyberthreats. It is not only the cybersecurity professionals who are leveraging the power of AI technology to prevent various types of cybersecurity threats. Even malicious actors are taking charge of the newest technologies to carry out cyberattacks on a large scale and in an automated way.

We are in the digital age where sensitive data protection and critical infrastructure immunity are the most important responsibilities of an organization. The integration of artificial intelligence has proved to be a transformative force in performing this operation. However, the traditional cybersecurity measures were quite efficient and did the job satisfactorily. They continuously faced several challenges with the ever-evolving tactics. This necessitated the use of AI and proactive and adaptive security measures. This chapter will help you understand two most important components in the threat detection and prevention system of cybersecurity:

1. Next-Generation Firewalls (NGFW) and Intrusion Detection Systems
2. AI-Powered Anti-Malware Solutions

The former section will delve into how the NGFW goes beyond rule-based filtering and utilizes the power of AI capabilities such as machine learning algorithms and behavioral analysis.

In the la⊚er part, we will explore the intricate world of modern malware threats from the subtleties of Advanced Persistent Threats (APTs) to the disruptive potential of ransomware and cryptojacking.

4.1 Next-Generation Firewall and Intrusion Detection Systems

4.1.1 Introduction to Next-Generation Firewalls

The world of cybersecurity is rapidly evolving, but the Next-Generation Firewalls have established themselves as an important component of defense against the innovative and sophisticated ways of modern cyberthreats.

Definition and Evolution

The Next-Generation Firewalls are a worthy upgrade to the traditional firewall, as they go many steps forward by incorporating advanced technologies to strengthen network security. At the core, the NGFW is an advanced security appliance that seamlessly blends traditional firewall functionalities with additional security features such as intrusion prevention, deep packet inspection, and application awareness. Unlike the old firewall system, the NGFW leverages a holistic approach. It scrutinizes the content and context of network traffic by going beyond mere ports and protocol filtering.

The traditional firewall, though efficient, was not capable of combating the new and complex ways of modern cybera◎acks surface. This led to the evolution of NGFW. The methods employed by adversaries are expanding and it becomes an absolute need of the hour to deploy an effective response. NGFW acts as an effective response to this paradigm shi◎ by introducing excellent security features to mitigate advanced cyberthreats.

Key Features of NGFW

NGFW is a part of the third-generation firewall technology which can be implemented in both hardware and so◎ware; making it capable of detecting and blocking any kind of a◎acks.

This firewall technology features advanced functions like:

- Deep Packet Inspection: To check the content of data packets that allows granular analysis of traffic and the identification of malicious payloads
- Intrusion Prevention System: Proactively detect and thwart potential threats including vulnerabilities and a◎ack pa◎erns
- Application Awareness and Control: Recognize and control applications traversing the network. A higher level of awareness enables be◎er control and enhanced security
- Threat Intelligence Integration: NGFW utilizes up-to-date threat intelligence feeds to stay ahead of emerging threats.
- Virtual Private Network (VPN) Support: Includes VPN support to ensure secure communications for remote users and branch offices.

In the following section, we will understand the role of AI in NGFW.

HOW NEXT-GENERATION FIREWALL WORK?

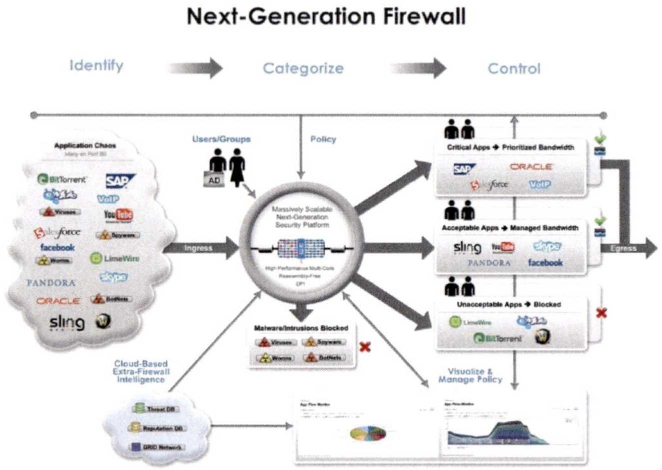

Source - https://www.linkedin.com/pulse/next-generation-firewall-ngfw-one-ossama-rashad/

4.1.2 Role of AI in Next-Generation Firewalls

No industry is left untouched by the influence of artificial intelligence. With the modern cybersecurity world continuing to face relentless challenges, it ought to leverage the power of AI in different cybersecurity operations. NGFW and AI focusing on machine learning algorithms, behavioral analysis, and the seamless integration of threat intelligence, have enhanced its effectiveness multiple times.

Machine Learning Algorithms

Machine learning algorithms power the artificial intelligence technology and AI-driven transformation in NGFW. It helps NGFW to adapt and learn from patterns within network traffic and distinguish between normal behavior and anomalous activities. Machine learning algorithms take help from historical data to continuously refine their understanding of the network environment. It helps NGFW identify upcoming threats, even those that manage to surpass traditional signature-based detection methods. Thus, it significantly enhances the NGFW's ability to combat sophisticated attacks in real-time.

Behavioral Analysis

Behavioral Analysis is an important component of AI-powered NGFW that complements the machine learning technique in cybersecurity. Behavioral analysis refers to observation and understanding of the behavior of users and applications within the network, rather than solely relying on predefined signatures or rules. So, whenever there is a deviation from the established norms, this technique flags it off as a potential threat and helps NGFW identify and mitigate zero-day attacks and novel tactics employed by cyberattackers. Behavioral Analysis makes NGFW highly adaptive and responsive; which is necessary in an environment where attack methodologies are in constant flux.

Threat Intelligence Integration

Integration of threat intelligence can improve the effectiveness of NGFW even further. AI-driven NGFW systems have access to a wealth of real-time threat data and incorporate feeds from global threat intelligence networks. So, this integration helps the NGFW with the latest information on different kinds of emerging cyberthreats, malicious entities, attack vectors, and vulnerabilities. By regularly updating itself with the latest information, the defenses are upgraded, fortifying the network against newer threats. So, threat intelligence integration not only helps identify and protect from known threats but also helps mitigate emerging and upcoming threats effectively.

4.1.3 Intrusion Detection Systems (IDS)

Intrusion Detection Systems (IDS) refer to the important component of cybersecurity; responsible for monitoring network traffic and devices for known malicious activity. It identifies and mitigates potential threats to network security.

How do Intrusion Detection Systems work?

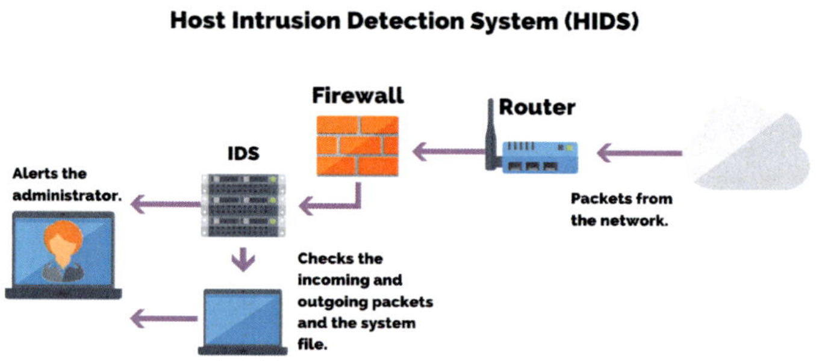

Source - https://www.liquidweb.com/blog/host-based-intrusion-detection-system/

Traditional IDS vs. AI-Powered IDS

Traditional IDS relied on existing signatures and rules to identify known pa◎erns and malicious activity within devices and networks. While the AI-powered IDS uses machine learning algorithms to analyze pa◎erns and identify anomalies and potential threats in real-time. Traditional IDS are effective in detecting established threats, but as technology is advancing and cyberthreats becoming more dynamic and sophisticated, they are struggling leading to the rise of AI-powered IDS. They go beyond the constraints of signature-based detection; and address the challenges posed by zero-day exploits.

It continuously learns to adapt to the nuances of evolving threats thus providing a proactive and adaptive defense mechanism; necessary in today's fast-paced cybersecurity world.

Real-Time Threat Monitoring

The greatest advantage of AI-powered IDS is its ability to monitor threats in real time. Traditional IDS operates on periodic scans, but the AI-powered IDS continuously monitors network activities, instantly identifies, and responds to potential threats as they come. This helps in a swift decisive response to malicious activities and reduces the window of vulnerability. It helps in minimizing the potential impact of diverse security incidents. By identifying anomalous behavior, IDS can mitigate the emerging threat, before it can escalate into full-scale security breaches.

Anomaly Detection

Anomaly detection is a major aspect of both traditional and AI-powered IDS. By incorporating anomaly detection into IDS, the security teams can identify subtle deviations that may indicate sophisticated attacks or insider threats. It can identify abnormal patterns of behavior, combined with real-time monitoring. It enhances the overall efficiency of IDS in maintaining the integrity and security of network infrastructures.

4.2 AI-Powered Anti-Malware Solutions

4.2.1 Understanding Modern Malware Threats

Malware is simply malicious software which is commonly known as computer viruses. As technology is evolving rapidly, so is the malware that is now capable of adapting and transforming to exploit vulnerabilities in digital ecosystems. In further sections, we will explore the evolution of malware, and highlight two insidious variants: Advanced Persistent Threats (APT) and the Pervasive threats of Ransomware and Cryptojacking.

Evolution of Malware

Malware has seen a relentless evolution over the years. They have evolved from simple viruses and worms to sophisticated threats; that can infiltrate an organization's digital infrastructure with high precision. This evolution reflects upon the advancement of cyberattackers, who continuously refine their skills and tactics to breach security systems. Once when malware was used to do nuisance-level disruptions; has now evolved into a highly advanced tool employed for financial gains, espionage, and various kinds of malicious activities.

Now, they include trojans, spyware, keyloggers, and polymorphic threats; that can change their code to evade detection. Therefore, it becomes necessary to understand this evolution for cybersecurity professionals; who want to develop effective countermeasures against evolving cybersecurity challenges posed by modern malware.

Advanced Persistent Threats (APTs)

Advanced Persistent Threats refer to advanced forms of malware that go beyond the traditional realm and demonstrate a high degree of sophistication. They are typically executed by well-funded and highly organized cybercriminals, who can be either nation, state, or cybercriminal groups with specific purposes. These types of threats can go unidentified for a long period and can pass through network defenses to collect confidential information or even execute targeted attacks.

Traditional malware is opportunist whereas the APTs are strategic and well-planned. These employ various techniques such as spear-phishing, zero-day exploits, and custom malware to achieve their goals. Thus, it becomes very difficult to detect and mitigate APTs as it requires advanced cybersecurity measures going beyond traditional antivirus solutions.

Ransomware and Cryptojacking

These are the two most prevalent and disruptive forms of modern malware. Ransomware refers to malicious software that encrypts a victim's files or systems; making them inaccessible. These files and systems can be unlocked only after the victim pays (usually cryptocurrency) to the attackers the ransom they demand. Ransomware attacks can target businesses, individuals, and even critical infrastructures that can cause significant losses and operational disruptions.

On the other hand, cryptojacking involves unauthorized use of the victim's computer systems to mine cryptocurrency. Cryptojacking malware is used to infect the system and then the hackers harness the processing power of compromised devices to mine cryptocurrencies such as Bitcoin or Monero. It not only affects the performance of the affected systems but also poses security and financial risks to the targeted individuals and organizations.

4.2.2 The Rise of AI in Anti-Malware

As the advancement and sophistication of malware have started affecting individuals and businesses, integration of advanced technology such as artificial intelligence has become necessary to build anti-malware solutions. In this section, we will understand how incorporating AI can help identify and mitigate malware attacks, especially using machine learning, behavioral analysis, and predictive modeling.

Machine Learning in Malware Detection

Including machine learning for threat detection indicates a paradigm shift in the world of cybersecurity and anti-malware solutions. While the traditional signature-based methods relied heavily on known patterns of malware, the machine learning algorithms help new anti-malware solutions to learn from vast datasets and recognize even subtle and evolving patterns that may indicate malicious behavior. Thus, helping the new systems to not only identify known threats but mitigate unknown variants as well.

Machine learning in malware detection goes beyond static analysis and enables the systems to adapt and evolve simultaneously with the ever-evolving tactics of cybercriminals. These AI-driven systems can continuously refine their understanding of normal and abnormal behaviors and provide a proactive defense against a wide array of malware.

Behavior Analysis and Heuristics

Behavioral analysis and heuristics have become an important aspect of AI in anti-malware solutions. The traditional methods struggled to identify polymorphic and behaviorally sophisticated malware. But with the use of AI-driven anti-malware solutions, this challenge can be overcome as they monitor the behavior of files and processes in real-time.

This technique involves assessment of the actions and interactions of software and identifying anomalies and malicious intent. Coupled with heuristics, which involves rule-based decision-making, this AI-driven anti-malware solution can identify patterns of abnormal behavior. This minute understanding helps to easily identify and contain the threats.

Predictive Modelling for Zero-Day Threats

Dealing with zero-day threats has always been the most critical challenge in the world of cybersecurity. It refers to the malware for which no known signature or defense mechanism exists at the time of discovery. AI plays a pivotal role in addressing this challenge through predictive modeling. They analyze historical data, threat intelligence feeds, and evolving tactics of cybercriminals to predict and defend against potential zero-day threats.

Predictive modeling goes beyond reactive measures and provides organizations with insights that can help them proactively strengthen their defense mechanisms. This in turn helps them reduce the window for vulnerability and allow for a resilient defense against diverse emerging and unknown threats.

4.2.3 Deep Learning Approaches

Deep learning is a subset of artificial intelligence that has revolutionized the field of anti-malware by introducing sophisticated neural network architectures capable of mimicking the human brain's ability to learn and recognize pa◎erns. In this part, we will explore the application of deep learning approaches in anti-malware through Neural Networks, Feature Extraction, and Pa◎ern Recognition and navigate through their advantages as well as challenges.

Neural Networks in Anti-Malware

Neural networks are inspired by the structure and functioning of the human brain. They are highly complex, and interconnected systems of algorithms. They are at the forefront of deep learning in anti-malware solutions. Neural networks can process vast amounts of data, making them suitable for the dynamic nature of malware detection.

TThey can identify pa◎erns that can indicate malicious activities by analyzing the features and behaviors of files and processes. They can be trained on diverse datasets to ensure they autonomously learn and distinguish between genuine and malicious entities.

Feature Extraction and Pattern Recognition

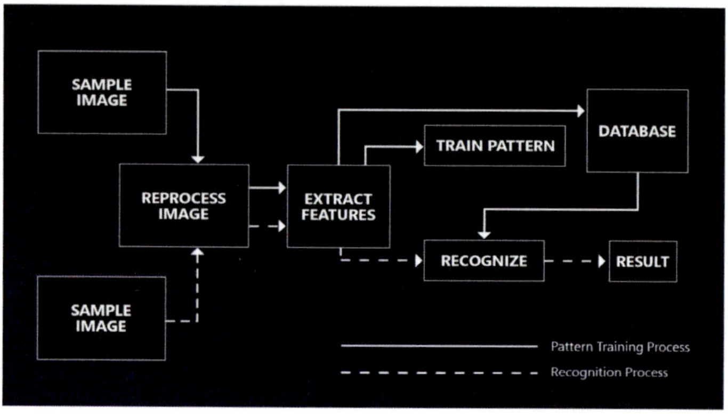

Automatic feature extraction and pa◎ern recognition are among the greatest strengths of deep learning techniques in anti-malware solutions. As discussed in the previous sections, the traditional anti-malware solutions o◎en relied on rule-based, and signatures, which made them less effective for newer threats. Deep learning, on the other hand, can autonomously identify relevant features and discern intricate pa◎erns with data. This makes them an accurate anti-malware detection mechanism.

Feature extraction involves identifying distinctive characteristics of files and processes associated with malware. Coupled with pa◎ern recognition, which entails understanding the relationships between these features, deep learning models can discern subtle nuances; indicative of malicious behavior.

Advantages of Deep Learning Approaches

- The deep learning models can adapt to new and unseen threats without manual intervention and provide a proactive defense against emerging malware.

- They can automatically extract intricate features and recognize complex patterns that enhance the accuracy of detecting sophisticated and polymorphic malware.
- They often result in fewer false positives, as compared to traditional methods because they can identify even subtle differences between malicious and normal behavior.

Challenges of Deep Learning Approaches

- The deep learning models require a huge amount of diverse and well-labeled data for their training. This becomes a challenging task to obtain the correct and accurate data; when it comes to the context of evolving malware threats.
- Since the deep learning models are highly complex, it makes them less interpretable, posing challenges in understanding the reasoning behind specific detections.
- The training and development of deep learning models require huge computational resources; making it difficult for resource-constrained organizations.

4.3 Continuous Threat Monitoring and Response with AI

Organizations usually follow the traditional cybersecurity methods, responding to security incidents, only after they have occurred. However, as the technological landscape is evolving, advanced methods of threat management are needed. With artificial intelligence, continuous threat monitoring and response becomes quite easier as it allows organizations a more robust security posture. In the following sections, we will understand the challenges related to traditional threat monitoring systems and how AI helps to address those challenges.

4.3.1 Challenges of Traditional Threat Monitoring

Traditional threat monitoring systems often come with various challenges that adversely impact their effectiveness in continuously monitoring and responding to threats. One of the significant challenges is their limited ability to evolve with the evolving cyberthreats. Cybercriminals use the most advanced methods to carry out their attacks and the traditional methods are often not ready for that. The traditional methods rely on static rules and signature-based detection methods, which sophisticated attackers can easily bypass using advanced techniques like polymorphic malware and zero-day exploits.

Not only that, traditional methods generate a lot of false positive alerts. This leads to overwhelming security teams. They find it difficult to identify the genuine threats. Traditional systems lack integration and interoperability features with modern IT infrastructure such as cloud services and IoT devices. This further affects their effectiveness.

Moreover, these traditional systems do not have proper analytics and machine learning capabilities. Such features are needed to identify the subtlest and most complex attack patterns. Therefore, traditional monitoring systems ultimately struggle to provide the real-time adaptive security, needed to guard against modern cyberthreats.

This sets the show clear as there is an increased vulnerability and risk for organizations.

4.3.2 How AI Enables Continuous Threat Monitoring

Artificial intelligence technology uses advanced algorithms and machine learning techniques to enable continuous threat monitoring, where they can detect and respond to threats in real-time. While traditional threat monitoring and response systems struggle to properly identify patterns and anomalies, advanced AI systems can easily analyze huge amounts of data to detect anomalies that can indicate potential cyberattacks.

These advanced systems can learn from past incidents and adapt to new and emerging threats. This infers that modern threat monitoring systems do evolve with the evolving cybercrimes.

Another advantage of integrating AI systems in cybersecurity is their sheer capability to process data at an unmatched speed. This helps identify threats immediately in real-time and mitigate them promptly.

False positives are greatly minimized by using AI for threat detection, thus enhancing accuracy, and helping security professionals to focus on genuine alerts. AI can incorporate behavioral analysis to detect insider threats or advanced persistent threats, that often go unnoticed by traditional threat monitoring methods.

Additionally, there are AI-powered automation tools that work round-the-clock and keep a vigil for any kind of lurking threats. This continuous, adaptive, and intelligent threat monitoring system strengthens an organization's defense against all kinds of new and emerging cyberthreats.

4.3.3 AI-Powered Threat Response

AI is not only about threat detection but enables organizations to promptly respond to cyberthreats in real-time.

- **Real-time Threat Detection and Response**

As AI systems can continuously monitor network activity, they can identify and respond to threats as they occur and help reduce potential damage.

Automated Threat Mitigation

AI automates responses for some of the common cyberthreats such as isolating compromised systems and blocking unnatural traffic. This automated system reduces the need for manual intervention and saves precious time in threat response.

Threat hunting and Investigation

AI systems can efficiently analyze data and identify the root cause of security incidents. This helps security professionals to understand the attack vectors and assists them in conducting investigations properly.

Predictive Security

As advanced AI systems can analyze historical data and threat intelligence, they can predict potential security threats and vulnerabilities even before they occur. This helps the security teams to take proactive measures to mitigate lurking and upcoming risks.

4.3.4 Benefits of Continuous Threat Monitoring with AI

Continuous threat monitoring and response with AI offers several benefits to organizations such as:

Strengthened Security Posture

AI can identify and respond to threats in real-time, this helps organizations provide a robust security posture and reduces cyberattack impact.

Reduces Response Time

AI systems automate responses to cyber incidents and prioritize alerts. This helps the security team to contain incidents faster and minimize damage.

Improved Efficiency

These systems can work round-the-clock with any fatigue and with maximum accuracy. They help security professionals focus on strategic work such as threat hunting and vulnerability assessment. Thus, it helps improve efficiency.

Continuous threat monitoring and response with AI helps organizations to stay ahead of the evolving cyberthreats. By integrating the latest AI tools and technologies, security teams can efficiently improve their threat monitoring and response.

CHAPTER 5

Machine Learning for Anomaly Detection

In the previous chapter, we learned how machine learning is powering the Next-Generation Firewall and Intrusion Detection System. It has proved to be a great technology for implementing anti-malware solutions. In this chapter, we will explore in detail how machine learning is serving as a great tool for anomaly detection as well. Here, we will uncover the subtleties of behavioral analysis in identifying suspicious activities and understand how to detect insider threats by leveraging the power of AI and machine learning.

5.1 Behavioral Analysis for Identifying Suspicious Activities

There are several strategies in cybersecurity to combat different types of cyberthreats and behavioral analysis is one of them. It scrutinizes the behavior of entities within a system and detects anomalies or deviations from the established norms.

It begins by establishing a baseline of normal behavior for users, applications, and systems, derived from historical data. Thereafter, there is continuous real-time monitoring aided by machine learning algorithms and statistical models. This allows deviation identification such as unusual login times or unexpected network communications. Let us have a deeper look at the behavioral analysis for identifying suspicious activities.

5.1.1 What is Anomaly?

In general terms, anomaly refers to something that deviates from normal. In cybersecurity, the anomaly can be understood as a rare occurrence or event, that is different from the standard pattern, and hence, can be deemed suspicious.

A few examples of anomalies can be:

- A transaction that does not fit with the user's previous purchasing style can detect credit card fraud
- Network traffic anomaly can indicate intrusion attempts, etc.

5.1.2 Types of Anomalies

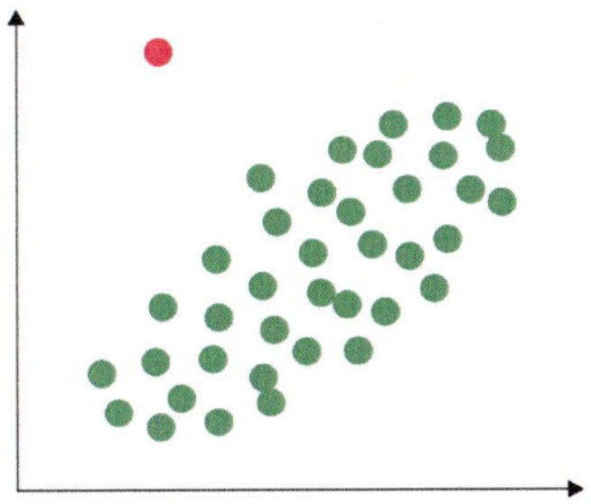

The red data point is a global outlier

Anomalies fall into three categories: Global Outliers, Contextual Outliers, and Collective Outliers. Here we shall understand the difference between each type.

Global Outliers

Global outliers, also known as Point Anomalies are the values that exist far outside the data set to which it belongs. They are simple and easy to identify as they are a rare event; and can be considered a global anomaly.

Example: You receive your salary each month in your salary account and one month, you suddenly receive millions of dollars. It can be easily identified as a Global Anomaly by the bank's analytics team.

Contextual Outliers

Contextual Outliers are data points that deviate significantly from the expected behavior within a specific context or subgroup. The "context" can be things such as time, location, or any other relevant factor. For instance, a data point might not be an outlier in the whole dataset, but it sticks out within a specific period.

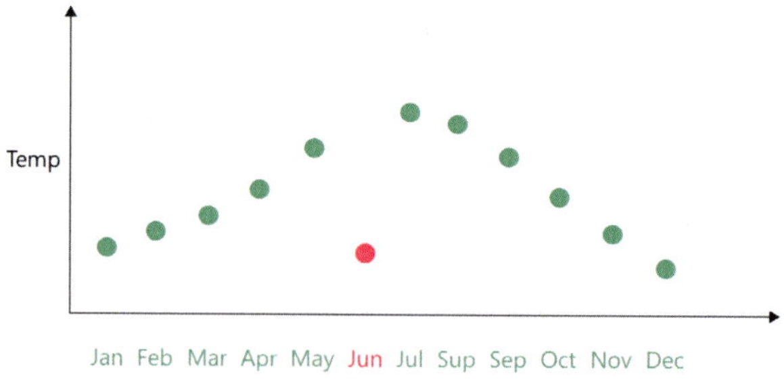

Example: The sales of fancy dresses rise during October. The sales are much higher than for the rest of the months of the year. But still, it is not considered an anomaly as it is expected that there will be a surge (as has been witnessed in previous years). However, if the sales are not as expected in October, then it can be considered an anomaly.

Collective Outliers

A Collective Outlier refers to a group of related data points that deviate significantly from the usual pattern observed in a dataset. Unlike individual outliers, which stand out automatically, collective outliers involve a set of interconnected observations exhibiting abnormal behavior together.

For example, in network traffic data, a collective outlier might occur when multiple devices within a network, display unusual communication patterns simultaneously. Detecting these anomalies is crucial for identifying coordinated or systemic issues that may indicate a security threat or irregularity in the overall system behavior.

5.1.3 What is Anomaly Detection?

Anomaly detection is a process of identifying unusual patterns or data points that differ significantly from standard and established behavior. The goal is to flag unusual occurrences that might indicate errors, defects, or security threats. In cybersecurity, anomaly detection plays an important role in identifying potential security breaches, malicious activities, or abnormal behaviors within a network or system. It involves analyzing data, often in real-time, to distinguish normal activities from unusual or suspicious ones. This is helpful for organizations to detect and respond to potential threats way before they damage the organization's data and infrastructure.

5.1.4 What is Behavioral Analysis and How is Machine Learning Used to Carry this Out?

Behavioral analysis is a technique that is used to analyze the patterns of behavior. It comes in handy in fraud detection and prevention by identifying and flagging incidents of fraud or other malicious activities. It involves several steps to perform this task:

i. Establishing a Baseline

The first step is to establish a baseline of normal behavior. This can be derived by studying the historical data and standardizing the expected pa◎erns of activities for users, applications, and systems.

Machine learning algorithms are trained with past organizational data that includes normal behavioral pa◎erns. This data helps the algorithm learn and establish a baseline that defines normal behavior.

ii. Continuous Monitoring in Real-time

The next step is continuous surveillance of activities. Behavioral analysis involves an ongoing real-time monitoring of activities by observing network traffic, user interactions, system processes, and other relevant data points.

iii. Detecting Anomalies

When there is continuous monitoring, the primary focus is to identify anomalies from the established baseline which can be unusual login times, unexpected network communications, or any other behavior that is not normal.

Trained machine learning models can identify anomalies by detecting deviations from the established normal behavior. This enables the system to flag unusual activities that may indicate potential security threats.

iv. Machine Learning and Statistical Models

Behavioral Analysis utilizes machine learning algorithms and statistical models that can adapt and learn from new data. This detects both known and unknown threats based on evolving pa◎erns and trends.

Machine learning models are adaptive and can continuously learn from new data. As the system encounters new patterns or behavioral shifts; the algorithms adjust while ensuring the model remains effective in identifying both known and emerging threats.

v. Contextual Analysis

Behavioral analysis considers the context behind different kinds of activities. For example, accessing certain confidential files can be normal for certain roles but unusual for others.

Machine learning aids in contextual analysis by considering various factors that influence behavior. By understanding the context surrounding activities, machine learning models enhance the accuracy of anomaly detection.

vi. Alerts and Response

When suspicious activities are detected, the behavioral analysis system generates alerts or notifications. These alerts highlight potential threats and require further investigation by cybersecurity teams.

vii. Incident Response

Behavioral analysis not only identifies anomalies but also facilitates a proactive response. Cybersecurity teams can take preventive measures, such as isolating affected systems, adjusting access permissions, or implementing additional security controls.

5.2 Detecting Insider Threats with AI

Insider threats have become more prevalent in organizations causing huge losses to businesses. Whether we talk about government organizations or businesses, this problem has been there since time immemorial. In modern-day businesses, critical information must be shared with different kinds of employees making it highly vulnerable, be it unintentional or malicious. These data can be compromised leading to severe and costly damages within organizations.

With technology evolving rapidly, malicious actors are devising new ways to breach security by remaining within the organization. This necessitates the use of more advanced techniques such as machine learning and artificial intelligence to detect insider threats. This section covers the details of what an insider threat is, what are its different types, and how AI can be leveraged to detect and prevent it.

5.2.1 What is an Insider Threat?

An organization does not only have to prepare its security strategy against cybercriminals acting from outside; but also, against employees who pose a potential risk of exposing the organization's confidential data to the outside world. An insider threat refers to the potential risk posed by individuals within an organization; who misuse their access, privileges, or knowledge to compromise the confidentiality, integrity, or availability of sensitive information, systems, or resources. Insider threats can come from employees, contractors, or business associates who, intentionally or unintentionally, act in ways that jeopardize the organization's security.

These threats can come in various forms, such as unauthorized access to confidential data, data exfiltration, sabotage, or the introduction of malware. For organizations, insider threats are one of the major concerns as they can bypass external security measures and exploit internal trust.

5.2.2 Types of Insider Threats

Insider threats come in various forms with each presenting a unique set of challenges to organizational security. Understanding these insider threats helps in effectively implementing mitigation strategies. Some of the common types of insider threats are:

1. Malicious Insiders

These are the individuals with insider knowledge who intentionally engage in harmful activities such as stealing data, sabotaging systems, or carrying out acts of revenge.

Example: An employee who intentionally poses damage to the organization by selling sensitive information he has access to, to the outsider.

2. Negligent Insiders

As the name suggests, negligent insiders are the professionals within the organization who do not follow or neglect the security protocols and procedures and inadvertently compromise the security due to their carelessness, and lack of awareness.

Example: An employee leaves open a sensitive document containing confidential information on a public computer.

3. Compromised Insiders

This refers to the set of insider threats whose credentials, despite all their efforts, are compromised by outside cybercriminals, thus making them an unwitting tool to carry out their cyberattacks from inside the organization.

Example: A phishing attack leads to the compromise of an employee's login credentials.

5.2.3 Different Ways AI Helps Detect and Prevent Insider Threats

Artificial intelligence and machine learning are great technologies that continuously monitor and analyze user behavior and their activities. AI-driven technologies can detect suspicious activities in the organization, identify patterns that can indicate potential threats, and respond proactively to mitigate the arising risks.

Below we shall discuss how AI can help detect and prevent insider threats:

Anomaly Detection

With AI-powered systems, organizations can create a normal standard of activities and behaviors based on historical data. It can serve as the baseline for every user, as well as systems. As explained in detail in the previous section, any deviation from this baseline can be easily detected with the help of machine learning algorithms and flagged as anomalies.

For example: If an employee tries to access data from a different department, say an employee from the marketing team tries to access data of the HR department, or if a login is detected at an unusual hour; say midnight. Then the AI system can trigger an alert, prompting further investigations, thus preventing huge damage.

Behavioral Analytics

AI systems are vivaciously capable of continuously monitoring and analyzing employee behavior and looking for pa◎erns that indicate risky or suspicious behavior. Behavioral analytics include analyzing employee's login pa◎erns, file access pa◎erns, data movement, and application usage. AI-driven systems can identify unusual behavior in real-time and detect potential insider threats; before they can cause any damage or losses to the organization.

Contextual Analysis

AI-powered systems can analyze the behavior of users in the context of their roles and responsibilities within the organization and check which sensitive data they are allowed to access. If there is any deviation from the norms established, then prompt action can be taken.

Privileged User Monitoring

Privileged users refer to system administrators or IT personnel who are responsible for maintaining the security of an organization's digital infrastructure. It is highly critical to monitor their activities to detect unauthorized or abnormal activities.

AI systems can monitor their activities closely and can check whether they are using their authoritative power to misuse their privileges or otherwise. Thus, a stringent check can be placed even on privileged users and administrators.

Data Access Controls

With AI systems, dynamic data access control can be implemented. It means the access rights can be adjusted based on the user behavior, and their responsibilities within the organization. The advanced AI systems can then easily detect any suspicious activity and automatically restrict access or even ask for additional authentication measures.

User Behavior Profiling

With the help of AI-driven systems, organizations can create detailed user behavior profiles. It considers different factors such as typical work pa◎erns, locations, and how applications are used. These AI systems can be◎er understand what the normal behavior is for each user, or system and efficiently detect deviations indicating insider threat.

Real-time Alerts and Response

AI-powered security systems are capable of continuous and real-time monitoring. They identify potential insider threats in real-time and send alerts to cybersecurity teams or administrators. This is an effective way to immediately investigate and proactively act to prevent further damage.

Just like the human brain, machine learning can continuously learn and adapt to the evolving insider threat pa⊚erns. This makes machine learning technology more effective than human-based security systems, for organizations. These algorithms can identify new trends and behaviors, allowing the system to refine its detection capabilities.

Chapter 6

AI in Cybersecurity Operations and Incident Response

AI in cybersecurity operations involves the utilization of machine learning algorithms and advanced analytics to analyze vast amounts of data in real-time. This enables cybersecurity professionals to identify patterns, anomalies, and potential threats that may usually be missed by traditional security measures. The ability of AI to continuously learn and adapt to new threat vectors makes it a powerful companion to fight against ongoing cyberthreats.

AI assists in automating the identification of security incidents, reducing response times, and providing valuable insights into the nature of the threat. By leveraging AI, security teams can prioritize and respond to incidents more effectively, allowing for a proactive and dynamic defense against cyberattacks.

This chapter delves deeper into how AI can be used to automate incident detection and response as well as how it can be used for forensics and investigation.

6.1 Automating Incident Detection and Response

In the previous chapters, we already learned how important it is to detect incidents on time and promptly respond to them so that the damage can be minimized. It does not ma◎er how strong the security system of any organization is, there is always room for improvement. Cybera◎ackers will explore every way to breach the security and at the times when the least number of professionals are active. Therefore, it becomes necessary to implement advanced automation technologies that can automate the incident detection and response; that can reinforce the organization's cybersecurity strength.

Using sophisticated algorithms and real-time data analysis, automated systems can prove to be very useful in identifying anomalies and potential security breaches. These in turn can help minimize response times and mitigate the impact of cyberthreats. In the further section, we will discuss in detail how AI technology can be leveraged to automate incident detection and response.

6.1.1 Reasons for Adapting AI-Based Incident Response Automation

Manual incident detection and response systems, however effective, were not feasible for evolving cyberthreats. Apart from that, several reasons require organizations to incorporate the latest automated threat detection and response systems. Here are some of the major reasons:

Remote work system: In the last few years, the remote working trend has grown larger than ever. Especially due to the impact of COVID-19, organizations found it suitable to work remotely. However, as the geographical distance increased because of remote work systems, the manual tracking of incidents in IT systems and networks became challenging. Remote work systems led to an increased number of data breaches. To handle such shortcomings, organizations understood the need for AI automation with intelligent threat detection.

Cost Effectiveness: It requires a lot of human resources, time, and cost in manual threat hunting. But with the implementation of AI, a one-time cost investment can bring efficient results for a long time to come making it more cost effective and easier.

Dynamic nature of Cybersecurity: A manual operation can only detect the threat when the a◉ack is initiated, or a long time has elapsed. This increases the chance of damage and losses. As the organizations wanted to resolve this problem promptly, they wanted a system that could identify threats and respond to them in real-time. Therefore, the need for AI automation has become more imperative.

6.1.2 How Automated Incident Detection and Response System Works?

The automated incident detection and response system leverages the power of AI and machine learning algorithms to identify and respond to different kinds of security incidents. Here is a general overview of how these systems work.

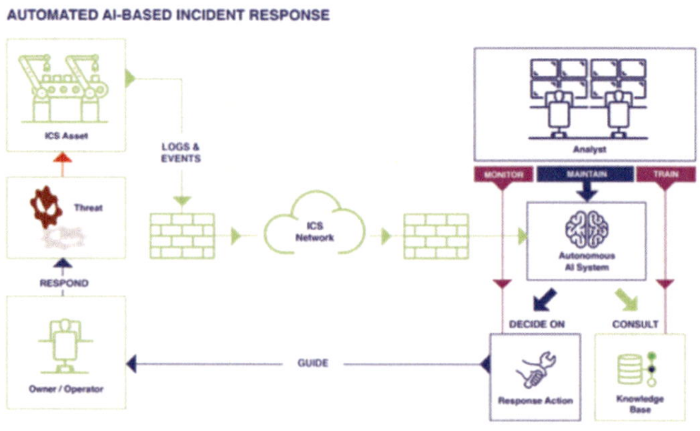

https://www.cyber.airbus.com/automating-the-soc-towards-ai-based-incident-response-in-the-factory-of-the-future-2/

Data Collection

Systems collect data from different sources including an organization's IT infrastructure consisting of server logs, network devices, applications, and other relevant sources.

Data Preprocessing

This step involves processing raw data i.e. cleaning and preparing it for analysis. It includes normalization, filtering, and handling missing values. This ensures accurate and relevant data is present for the subsequent machine learning algorithms.

Feature Extraction

It refers to the process of extracting relevant features or attributes from data that are already processed. These features help set input variables for the machine learning models. For example, features can be used to define login patterns, file access, network traffic, etc.

Machine Learning Algorithms

The automated incident detection and response system employs machine learning algorithms to analyze historical data and define patterns, anomalies, or deviations from the established norms. Supervised learning, unsupervised learning, and anomaly detection algorithms are some of the common machine learning techniques used for this purpose.

Training the Model

Then the machine learning models are trained using the prepared datasets. They are trained about which behaviors are normal and what could be malicious. During this phase, the models learn to recognize patterns associated with security incidents.

Real-Time Monitoring

The models are trained and deployed a◎er which the system continuously monitors the incoming data in real-time. The system compares the current activities with the trained data and pa◎erns learned.

Automated Incident Detection

When the system finds out that there is a deviation from the established norms, it can identify and label it as an anomaly or suspicious activity. The system activates a trigger that sends an alert and notifies the concerned professional from the security team. The alert can be the details about the nature of the incident, the affected systems, or the potential severity.

Automated Response

Some systems can be advanced, and depending on the incident, a predefined response strategy can be implemented. It includes isolating affected systems, blocking suspicious network traffic, or any other actions predefined to mitigate the impact of the incident.

6.1.3 Benefits of Automating the Incident Detection and Response

Automating incident detection and response offers numerous benefits for organizations aiming to enhance their cybersecurity strategies. Some key advantages are:

- Automated systems can swi◎ly analyze large volumes of data in real-time, allowing for the rapid identification of security incidents.
- They can operate continuously, providing around-the-clock monitoring and vigilance.

- Such mechanisms can immediately execute predefined actions when a threat is detected. This reduces the time between detection and response, minimizing the window of opportunity for attackers to exploit vulnerabilities.
- It can automate routine and repetitive tasks to help organizations optimize the use of human resources. Security teams can focus on more complex and strategic aspects of incident response, while automated systems handle routine actions.
- Automated systems can be updated and retrained to adapt to new and emerging threats. Machine learning models can learn from the latest data, allowing the system to evolve and improve its ability to detect and respond to new attack tactics.

6.1.4 Automated Incident Response Tools

Some open-source tools include:

- TheHive
- Security Onion
- OSSEC
- OpenDLP
- Elastic Stack
- Shuffler
- Automated Incident Response and Forensics Framework for AWS

6.1.5 The Need for Human Intervention and Collaboration

The automated incident detection and response system is a great way to automate the entire detection and response process, but it must be noted that they are not a silver bullet. When it comes to the security of an organization and its digital assets, human intervention and collaboration with these AI systems are very crucial. Human expertise and oversight will always remain essential in making complex decisions, conducting thorough investigations, and ensuring the effective performance of these systems.

6.2 Using AI for Forensics and Investigation

The integration of AI into digital forensics offers a ton of advantages, with automation being a key area of enhancement. Traditional digital forensics processes are o◎en laborious, time-consuming, and prone to human errors. AI technologies can automate the identification and classification of evidential data, streamlining the entire process, and significantly enhancing accuracy.

AI excels at efficiently handling 'big data' as well, surpassing the capabilities of human investigators. The rise of IoT devices and cloud technology has led to an unprecedented increase in daily data generation, along with an increased cybercrime rate. AI-driven digital forensics can rapidly process vast volumes of data, uncovering meaningful pa◎erns, trends, and anomalies with speed and efficiency.

Adding to this, the adaptability and continuous learning inherent in machine learning algorithms play a pivotal role. AI systems not only benefit from each investigation but also evolve, improving their performance and adapting to emerging forms of cybercrimes.

6.2.1 What are Digital Forensics?

Digital forensics, is a branch of forensic science, and is also known as digital forensic science or cyber forensics. It deals with the collection, analysis, and preservation of electronic evidence to investigate and prevent cybercrimes or other forms of digital incidents. Digital forensics and investigation aim to uncover and examine digital evidence from various electronic devices and digital media to support legal proceedings.

Key components of digital forensics and investigation include:

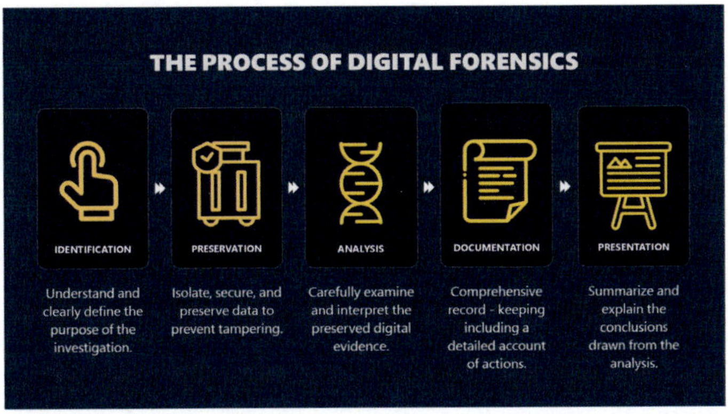

Evidence Collection

Gathering electronic evidence from a variety of sources, including computers, mobile devices, storage media, networks, and cloud services. This involves ensuring the integrity and authenticity of the collected data.

Analysis of Digital Evidence

Examining digital evidence to uncover relevant information, pa◎erns, and insights. This process utilizes specialized tools and techniques for extraction, interpreting, and analyzing data from different types of digital media.

Preservation of Evidence

It means ensuring the integrity where digital evidence is preserved properly so that it remains admissible during the legal proceedings. This involves documenting the chain of custody and using proper procedures to prevent tampering or alteration.

Recovery of Lost or Deleted Data

Employing techniques to recover lost, deleted, or hidden data from digital devices. This is crucial for reconstructing events and understanding the actions taken by individuals involved in cybercrimes.

Incident Response

Responding to and managing cybersecurity incidents, including data breaches, malware a◎acks, and other digital threats. Digital forensics is an essential component of incident response, helping organizations understand the nature and extent of the incident.

Network Forensics

In this, the network traffic is investigated and analyzed to identify if there is any kind of security incident, or unauthorized access is happening, or if any kind of other network-related issues are present. Network forensics helps trace the activities of individuals within a network environment.

Forensic Imaging

Creating a forensic image or copy of digital storage media to preserve the original state of the data. This image is then used for analysis, ensuring that the original evidence remains unaltered.

6.2.2 The Need for AI in Digital Forensics

One of the biggest challenges in digital forensics is data overload. Mountains of files – emails, texts, social media posts, browser history – hold critical clues, but manually combing through them is akin to searching for a needle in a haystack. Thus, there was an absolute need to automate the digital forensics that could perform this task rapidly without any error.

AI-powered solutions serve as a game changer by employing machine learning algorithms. These algorithms can autonomously categorize and prioritize files based on relevance, file type, timestamps, and even keywords. They are capable of instantly extracting all financial transactions from terabytes of data, significantly narrowing down the search for stolen assets. This saves investigators precious time and allows them to focus on the most promising leads.

6.2.3 How is AI Used for Revolutionizing Digital Forensics?

There are different ways in which AI is used for digital forensics.

Unmasking the Tactics

Malicious actors employ sophisticated techniques to mask their activities, erase tracks, and blend in with the noise. But AI can help detect these intricate malicious activities and unmask their attempts. Natural Language Processing (NLP) algorithms can analyze text for suspicious language patterns, emotional cues, and inconsistencies. AI can identify bot accounts; manipulate online discourse and pinpoint hidden messages embedded in seemingly normal files. AI can even flag anomalies in network traffic, to reveal covert data exfiltration attempts.

Image and Video Analysis

Visual evidence plays a crucial role in investigations, and AI is unlocking its full potential. Image and video analysis algorithms can scan through hours of footage, automatically recognizing faces, objects, and activities. AI can swiftly identify a suspect in a crowded surveillance video and point out when a critical file was accessed on a computer screen. AI can even reconstruct deleted images or videos by piecing together fragmented clues to paint a clearer picture of events.

Forecasting Emerging Threats

Forensics is not just about the past; it is about anticipating the future. AI can leverage predictive analytics to forecast potential cyberattacks, data breaches, and even criminal activity. By analyzing past incidents, identifying patterns, and learning from them, AI can alert investigators about looming threats, allowing them to proactively take preventive measures. It can predict future phishing attacks based on historical campaigns or identify individuals with a high risk of engaging in cybercrime. This proactive approach is helpful in significantly reducing the impact of criminal activity and improving overall security.

6.2.4 Applications of AI in Digital Forensics and Investigation

AI has a wide array of applications when it comes to Digital forensics and investigation.

1. Email Forensics

NLP for Analyzing Text: Identify suspicious language patterns, emotions, and keywords in emails to pinpoint phishing attempts, malware scams, and insider threats.

Social Network Analysis: Trace email chains and connections between individuals to map out communication networks and uncover potential conspiracies.

Sentiment Analysis: Gauge the emotional tone of emails to understand intent, motives, and potential red flags for criminal activity.

2. Image and Video Forensics

Facial Recognition: Identify suspects, witnesses, and persons of interest in surveillance footage and video recordings.

Object Detection: Locate specific objects (weapons, drugs, stolen items) within images and videos, accelerating evidence search and analysis.

Deepfakes Detection: Unmask synthetically generated videos and images used for disinformation or fraud.

3. Network Forensics

Anomaly Detection: Identify unusual traffic patterns, unauthorized login attempts, and potential data breaches based on network activity deviations.

Malware Identification: Analyze network packets to discover malicious code in transit and track its source and destination.

Predictive Modeling: Forecast potential cyberattacks based on historical network security incidents and threat patterns.

4. Mobile Forensics:

App Analysis: Identify malicious apps, track data usage, and uncover hidden files and activities on mobile devices.

Call and Text Analysis: Extract and analyze call logs, text messages, and voicemail records to reveal communication pa◎erns and connections.

Location Tracking: Analyze GPS data and cell tower pings to reconstruct a timeline of a suspect's movements.

5. Cloud Forensics:

Data Extraction and Classification: Efficiently extract and categorize vast amounts of data stored in cloud platforms like Dropbox, Google Drive, and Microso◎ OneDrive.

Deleted Data Recovery: Recover deleted files and emails from cloud storage, potentially revealing hidden evidence.

User Activity Monitoring: Track user activity logs and access pa◎erns within cloud platforms to identify unauthorized access and data breaches.

6.2.5 Challenges of Implementing AI

It is not all sunshine and rainbows. AI in forensics comes with its own set of challenges. One major concern is bias. AI algorithms are only as good as the data they are trained on. If they are trained on biased data, then it can lead to skewed results and potentially incorrect conclusions. To ensure fairness and accuracy, rigorous training and testing of AI models are important.

Another concern is transparency. It is hard to understand how complex AI algorithms conclude, as their workings are o◎en opaque and difficult to understand. For legal proceedings, establishing a clear chain of evidence and explainability will be critical.

Despite all these challenges, AI has a huge potential in digital forensics. As technology matures and ethical considerations are addressed, AI will become an indispensable tool for investigators, ushering in a new era of speed, precision, and proactivity in the fight against digital crime. The future of forensics lies not in replacing human investigators, but in empowering them with AI's superpowers, so they can crack the most complex digital cases and bring the most cunning criminals under the radar.

6.3 AI-Powered Threat Hunting and Vulnerability Management

For a long time, threat hunting and vulnerability management was a manual process that relied on the expertise of security analysts. But as the amount and technology used to carry out attacks is expanding; traditional methods are no longer effective, and it requires a more efficient and scalable approach. Artificial intelligence fulfills this demand by augmenting human capabilities and providing a more proactive approach to cybersecurity.

This chapter focuses on the challenges organizations used to face with traditional threat hunting and vulnerability management, and how AI has completely transformed this process.

6.3.1 Challenges of Traditional Threat Hunting and Vulnerability Management

Some of the most persistent challenges of traditional threat-hunting and vulnerability management methods are:

Time Constraints

Needless to mention how time-consuming and resource-intensive the manual searching for hidden threats within large datasets is. So, security analysts must strive to evolve with the changing threat environment.

Limited Visibility

As technology is ge◎ing advanced, the threats are also becoming highly sophisticated employing entirely new techniques and exploiting unknown vulnerabilities. Such a◎acks are hard to detect with traditional tools.

Reactive Approach

Traditional methods are good only when the a◎ack has happened as they allow security teams to react a◎er security incidents; which is beneficial for a◎ackers to exploit in that period.

6.3.2 Empowering Threat Hunting with AI

With conventional methods struggling to manage vulnerabilities, AI offers great advantages to enhance threat hunting.

Automated Threat Detection

AI algorithms can analyze huge amounts of security data related to network traffic, user behavior, endpoint activities, and more. This helps with continuous threat hunting and assists security teams in identifying subtle anomalies that can indicate malicious activity.

Advanced Analytics

Artificial intelligence leverages new techniques like User Entity and Behavior Analytics (UEBA) to identify suspicious pa◎erns or anomalies that deviate from set standard behavior and helps detect insider threats and compromised accounts easily.

Prioritization and Alert Filtering

AI excels in prioritizing alerts based on several factors like the severity of threats, and their contexts, thus helping security analysts to prioritize and address the most critical threats.

6.3.3 AI-Powered Vulnerability Management

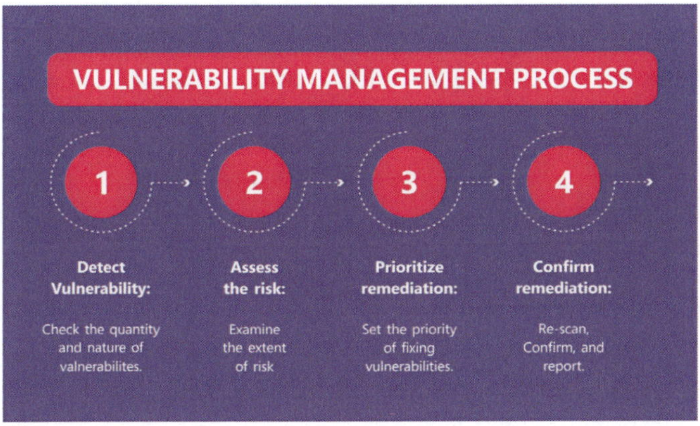

Just like effective threat hunting, AI makes vulnerability management a cool breeze. Here is how:

Automated Vulnerability Scanning and Prioritization

AI's capability to automate vulnerability scanning processes across different systems makes it highly superior. It can also prioritize the vulnerabilities based on how easy they are to exploit, how they could impact the organization's business operations, and how critical the asset is. So, security teams can know which one to patch first.

Vulnerability Correlation and Risk Assessment

Another important aspect of integrating AI for vulnerability management is it can easily correlate vulnerabilities across different systems and identify the attack vectors that combine multiple vulnerabilities. Thus, security teams can prioritize the preventive measures depending on the overall risk the vulnerabilities pose to systems.

Predictive Risk Analysis

AI assists in predicting future vulnerabilities and potential risks by analyzing past data and threat intelligence. This helps security professionals take appropriate steps to mitigate the attack even before the attackers try to exploit available vulnerabilities.

In conclusion, AI is a boon for threat hunting and vulnerability management processes and helps organizations to effectively identify risks beforehand and take preventive measures even before they occur. By using AI's power to analyze huge datasets, and automate necessary tasks, organizations can maintain a better and robust security posture.

Chapter 7

The Ethical Implications of AI in Cybersecurity

In the previous chapters, we learned how the implementation of AI has been proven to revolutionize the cybersecurity landscape. From threat detection to implementing prevention strategies, AI can automate the entire cybersecurity process with the least human intervention. However, this integration brings forth many ethical implications that need careful consideration. In this chapter, we will look closer into the intricate relationship between AI and cybersecurity and explore the ethical dimensions that arise from the deployment of intelligence systems in digital defense.

Here we will explore the ethical responsibility associated with granting machines the authority to make critical decisions regarding network security. AI algorithms are evolving to autonomously identify and neutralize potential threats; therefore, questions arise about their accountability and transparency. It becomes indispensable to examine the ethical frameworks that guide the development and implementation of AI in cybersecurity.

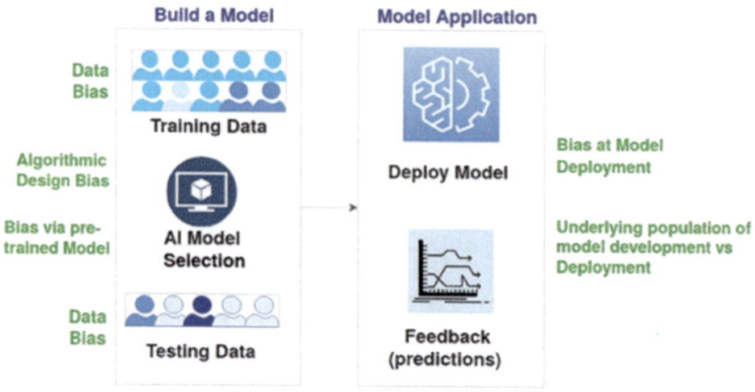

7.1 Bias and Fairness in AI Algorithms

As the use of AI in the field of cybersecurity has been increasing rapidly, it becomes very important to gauge the inherent potential for bias and fairness in AI algorithms. In the following sections, we will see the impact and mitigation of "Bias and Fairness in AI algorithms" in the context of cybersecurity. Bias can have a subtle influence in shaping the decision-making processes of AI systems, whereas fairness ensures just and equitable outcomes.

7.1.1 Types of Bias in AI Algorithms

Before we dive deep, it is wise to understand different types of bias that can affect decision-making in the cybersecurity space.

- **Algorithmic Bias**

Algorithmic bias refers to the biases introduced during the design and development of algorithms arising from the underlying assumptions and choices made by developers.

For example: If a cybersecurity algorithm is trained on historical data that reflects a disproportionate focus on certain types of threats, the algorithm may develop biases toward those threats, potentially overlooking or underestimating others.

Data Bias

Data bias arises when training datasets used to teach AI algorithms contain imbalances or reflect historical prejudices, leading the algorithm to perpetuate and even amplify existing biases.

For example: If a cybersecurity AI is trained on data that predominantly features a◎acks from specific regions or industries, it may exhibit a bias in threat detection, potentially overlooking emerging threats from underrepresented areas.

Representation Bias

Representation bias occurs when certain demographic or identity groups are underrepresented or overrepresented in the training data, leading the AI to exhibit skewed behavior toward those groups.

For example: If the training data for a cybersecurity algorithm predominantly comprises a◎acks on a specific industry, the algorithm may struggle to effectively detect threats targeting industries that are less represented in the data.

Socio-economic Bias

Socio-economic bias emerges when AI algorithms inadvertently favor or disadvantage individuals based on socioeconomic factors, such as income, education, or geographic location.

For example: A cybersecurity algorithm may exhibit socioeconomic bias if it primarily focuses on threats to high-profile organizations, potentially neglecting the security needs of smaller businesses or individuals with fewer resources.

7.1.2 Consequences of Biases

If there are biases in the algorithms, it can result in a severe impact on the world of cybersecurity. It will lead to unintentional as well as destructive outcomes that can severely impact organizations, individuals, and even society.

Here let us explore some of the major consequences that can arise because of algorithmic biases:

- **Discriminatory Outcomes**

Algorithmic bias can result in discriminatory outcomes where certain individuals, communities, or groups can be unfairly targeted or can be affected by cybersecurity measures.

If there is bias in the threat detection algorithm towards a specific demographic or industry, then there are chances that it will neglect risks over those biased communities and leave them more vulnerable to cyberattacks.

- **Uneven Resource Allocation**

If there is bias in the algorithm, then it will direct more attention and resources to certain sectors or types of threats which will lead to uneven allocation of cybersecurity resources and can severely impact the areas that need more attention.

For example, if the algorithm is biased to prioritize threats against high-profile organizations only, then it will allocate only a small number of resources to protect organizations that are not on priority.

- **Affects trust in the AI-driven Cybersecurity world**

Once the stakeholders start questioning the legitimacy of the algorithms used in the AI-powered cybersecurity system, there are chances the trust will be eroded.

If organizations or individuals lose confidence and perceive that the cybersecurity AI is preferring certain groups and leaving others, then they will lose confidence in the system's ability to provide unbiased decisions.

Exaggerate Existing Societal Inequities

AI can exaggerate the existing societal inequalities by reinforcing and perpetuating the biases present in training data.

Cybersecurity algorithms can heavily amplify the disparities if the historical biases in the training data are not properly addressed. This leads to certain communities being left out and more exposed to risks.

Compromised Effectiveness in Threat Detection

Algorithms can fail to accurately identify and respond to different kinds of cyberthreats because of bias. It may not recognize familiar attack patterns leaving the systems vulnerable to emerging threats.

Legal and Ethical Consequences

A biased algorithm can also lead to legal and ethical challenges which can result in lawsuits, regulatory scrutiny, and even reputational damage to organizations deploying affected systems.

7.1.3 Challenges to Identifying and Mitigating Bias in Algorithms

In the context of cybersecurity, identification, and mitigation of bias in algorithms is very much necessary but quite challenging as well. These challenges can be technical, ethical, or practical. Organizations must recognize and address these challenges properly to develop an effective cybersecurity strategy that will ensure fairness in AI-driven cybersecurity efforts.

Some of the challenges that need to be addressed include:

Opaque Algorithmic Processes

Many AI algorithms, particularly complex ones employed in cybersecurity, operate as black boxes, making it challenging to figure out their decision-making processes. This lack of transparency affects the efforts to identify and understand the sources of bias introducing obstacles in the development of an effective mitigation strategy.

Evolving Form of Cyberthreats

The cybersecurity world is highly dynamic where new and varied forms of cyberthreats are employed to continuously exploit organizations and individuals. AI algorithms must therefore adapt continuously to evolving cyberthreats but if there are biases in datasets, then many of those threats will go unnoticed. Outdated data will also lead to algorithms becoming ineffective in recognizing and mitigating cyberthreats.

Inaccurate or Incorrect Datasets

One of the biggest challenges in making an effective AI-powered cybersecurity model is building comprehensive and unbiased training datasets that can accurately reflect different types of cyberthreats. If the data is inaccurate or incomplete, then it can introduce bias which will further lead to underperformance of the threat detection and response systems.

Subjectivity in the Definition of Fairness

Fairness can be considered as a subjective concept. Different stakeholders may have different perspectives about what is fair and what is not in the context of cybersecurity. Therefore, this disagreement in fairness and its definition can make the process of developing a universally acceptable mitigation strategy complicated.

Limited Explainability of AI Decisions

Many AI algorithms lack proper explanation making it challenging to properly analyze why a particular decision or prediction the model has made. If there is no clear explanation, then stakeholders might struggle to understand and address the bias which can prove to be an obstacle for mitigating efforts.

Continuous Monitoring and Adoption

Over time, biases can emerge in AI algorithms that require continuous monitoring and adaptation to evolving conditions. If continuous monitoring is ignored, then it can lead to the persistence of bias compromising the overall effectiveness of cybersecurity measures.

7.1.4 Regulations to Monitor Bias in Algorithms

Now policymakers have started recognizing the need to address ethical concerns and biases related to the integration of AI in cybersecurity and this is why the regulatory landscape has been rapidly evolving. There are several jurisdictions, regulatory bodies, and organizations that are formulating guidelines and standards to govern the development, deployment, and use of AI algorithms in the cybersecurity world.

These regulations ensure a proper transparency, accountability, and fairness are maintained in the AI systems, especially concerns related to the potential biases that can impact cybersecurity outcomes. On top of that, data protection and privacy regulations are also contributing positively to shaping how AI algorithms handle and process sensitive user data.

As we move ahead, the regulatory environment will mature further and organizations that are engaged in cybersecurity efforts will have to align their security operations with the regulatory standards. This will help promote responsible and ethical use of AI to protect digital ecosystems.

7.1.5 Mitigating Bias in AI Algorithms

The most inevitable thing is how to mitigate biases in AI algorithms. However, it is challenging to completely curb the bias, strategic care can be taken to minimize the risk these biases bring in cybersecurity operations.

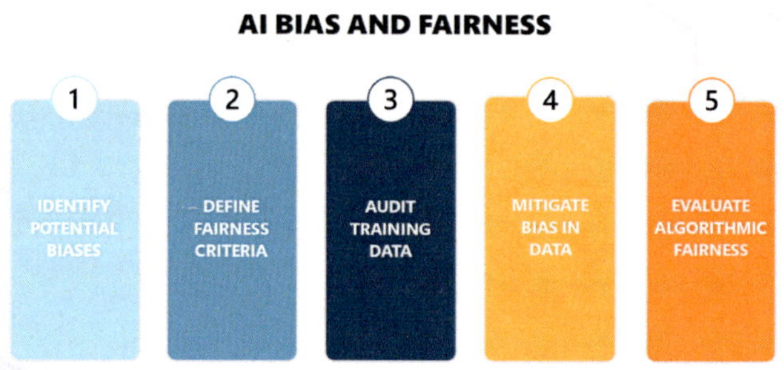

Here are some things that must be checked:

- Ensuring the training datasets are complete, accurate, and diverse. It must include a wide spectrum of cyberthreats and potential targets.
- Transparency in the AI algorithms should be prioritized which can make decision-making processes more legitimate and understandable.
- There should be regular monitoring and auditing to ensure there are no biases in the algorithms. If identified, they must be rectified instantly.
- During the development and deployment stages, defining and incorporating fairness metrics can benefit substantially. Regular assessment of the performance of the AI algorithms using these metrics should be done that ensure outcomes align with the predefined fairness standards.

- Integrating ethical design principles into the development process and emphasizing fairness, transparency, and accountability
- Utilizing specialized tools that are designed to detect and measure biases in AI algorithms during the development phase can address biases in different stages of model creation.

These mitigation strategies are a great way to minimize biases and effectively integrate AI into cybersecurity measures. Organizations following these mitigation strategies can ensure AI algorithms are also used positively without causing any adverse effects.

7.2 Privacy Concerns and Data Protection

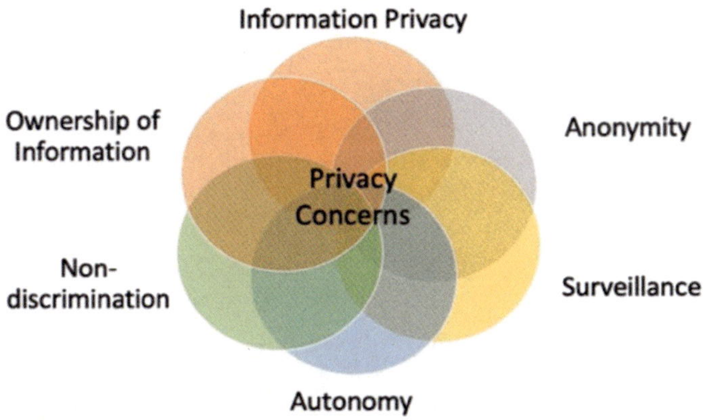

https://openeducationalberta.ca/educationaltechnologyethics/chapter/ethical-considerations-when-using-artificial-intelligence-based-assistive-technologies-in-education/

Furthermore, privacy concerns and data protection are critical issues that cannot be overlooked. AI algorithms have proven to be a great tool in protecting an organization's digital assets and infrastructure from various kinds of cyberthreats. Balancing the efficiency of these AI systems along with the protection of individual policy has become a significant challenge to address.

As we move further in this section, we shall examine the intricate relationship between privacy and cybersecurity and explore the ethical challenges that may arise due to the collection, processing, and utilization of sensitive data in AI systems.

7.2.1 The Interplay between Privacy and Cybersecurity

It is important to understand the relationship between privacy and cybersecurity when we discuss about ethical considerations of using AI in cybersecurity for threat detection and prevention. Although, technologically we have advanced beyond measure, still there is a huge gap between strengthening cybersecurity measures and protecting individual privacy rights. To maintain a balance between advanced technologies like AI algorithms and protecting digital ecosystems, it is necessary to manage these things properly.

The foremost thing organizations need to do is explore the ethical landscape; when they are collecting, analyzing, or utilizing data in any form. Undoubtedly, data is the fuel that powers the advanced AI systems and it is the data only that helps to identify and mitigate cyberthreats. But this data can consist of sensitive information which can raise questions regarding the ethical handling of personal and confidential data. Thus, it becomes even more important to maintain an equilibrium where cybersecurity measures are enhanced without affecting the privacy of individuals.

Furthermore, fair, and transparent data practices are key including clear communication about why and how data is collected, how they can be used, etc. The users must give their consent to share their information. As AI algorithms are evolving with time and adapting to emerging threats, ethical considerations will ensure the responsible implementation of these technologies to protect individual rights and privacy and at the same time combat threats effectively.

7.2.2 Privacy Risks of Using AI Systems

AI and Intrusive Surveillance

Invasive monitoring is one of the biggest concerns related to the AI's capability of enhanced surveillance systems. AI can prove to be an excellent tool for facial recognition, biometric analysis, and predictive algorithms, that can track the movement of individuals in online as well as offline spaces and scrutinize them for analysis. This increased surveillance can pose a risk to privacy as it can encroach on personal spaces and can lead to unwarranted intrusion into private lives.

Profiling and Predicting Future Behavior

AI can easily analyze a vast dataset to create intricate profiles and predict individual's behavior and preferences. With the help of machine learning algorithms, it can identify pa◎erns and make predictions about future events depending on historical data. Though it sounds astounding and will be highly beneficial in several cases, it also raises concerns about its potential misuse. Predictive profiling has a greater chance of leading to privacy infringement by making intrusive assumptions about individual's personal lives.

Discrimination and Manipulation

Algorithmic bias is inherent in the training data which brings high risks to the use of AI. If the training data contains biases, AI algorithms can make incorrect decisions and even amplify the existing biases which will result in discriminatory outcomes.

It will further reinforce the existing societal disparities because of the unfair treatment of sections, groups, or communities. On top of that, AI-generated data analysis can also be exploited for manipulative purposes which can influence an individual's behavior and decisions through targeted content, advertisements, or misinformation campaigns, compromising autonomy.

- **Privacy Fatigue**

The constant collection and analysis of personal data contributes to a phenomenon known as "privacy fatigue." As individuals experience repeated invasions into their privacy, a sense of helplessness and resignation may set in. The amount of data being collected is rising rapidly and when it is combined with complex AI-driven analyses, it can heavily impact individual's sense of control of their personal information. Privacy fatigue indicates the frustration of individuals who feel overwhelmed by their continuous tracking. It also includes the profile that is affecting their autonomy and leading to a reduced enthusiasm for participating in privacy protection measures.

7.2.3 Potential for Privacy Breaches and Data Flaws

There is a greater chance of privacy breaches and data leaks when there is a convergence between extensive data and highly complex AI models. AI can analyze diverse datasets which makes it a valuable target for exploitation by malicious actors. If there is any unauthorized access or data leaks, then it could have very high adverse consequences for both individuals and organizations. It might include identity theft, financial fraud, or reputational damages as well. If we look at the societal level, large-scale data breaches can affect public trust, impact economic stability, and lead to widespread social consequences.

7.2.4 Limitations of Current Data Protection Laws

Though there are several data protection laws in place, there are certain limitations that prevent them from effectively addressing various kinds of challenges posed by AI. The first challenge is the rapid evolution of technology and the traditional regulatory frameworks struggle to match their pace resulting in specificity in regulating AI applications. AI-driven data processing is highly complicated and includes algorithmic decision-making and machine learning which pose a significant challenge to the legal frameworks that are designed for traditional data processing methods. There can be confusion in defining the accountability, responsibility, and complexity of AI-related data processing which can affect the effectiveness of current regulations.

7.2.5 Different Data Protection and Privacy Regulations

Currently, there are several data protection and privacy laws employed in different parts of the world with each country implementing its regulations in the best interest of its citizens. Some of the most popular laws from around the world have been mentioned below:

Europe

General Data Protection Regulation (GDPR): This is arguably the most comprehensive and detailed data protection law in the world. It applies to all the companies processing the personal data of EU citizens; no ma◎er where the companies are located. This law helps to empower individuals with a wide range of rights over their data such as the right to access, rectify, erase, and restrict processing.

ePrivacy Directive: This is another major regulatory framework in Europe that focuses on the privacy of electronic communications such as emails, phone calls, messengers, etc. It regulates cookies, website tracking, and the use of metadata.

United States

California Consumer Privacy Act (CCPA): Similar to GDPR in Europe, this law in the US grants the residents of California similar rights over data including the right to know what information is collected, to delete it, and to opt out of its sale.

Children's Online Privacy Protection Act (COPPA): This law is designed to protect children under 13 and requires parental consent for the collection of their personal information.

Other Notable Laws

Brazil's General Data Protection Law (LGPD): LGPD is designed for Brazilian citizens and grants them control over their data and requires organizations to ensure they process their user's data lawfully and securely.

Canada's Personal Information Protection and Electronic Documents Act (PIPEDA): This law protects the privacy of Canadians' personal information collected by commercial organizations.

China's Personal Information Protection Law (PIPL): This law from China regulates the collection, use, storage, and processing of personal information by organizations all over the world that are operating in China.

As we stride ahead into the future, we will witness an increased use of artificial intelligence in cybersecurity. It is an unavoidable fact that AI will bring some severe ethical challenges with it that cannot be ignored. It is time that organizations, governments, and policymakers perform a collective effort and design regulatory frameworks for ethically implementing AI in the world of cybersecurity that will not only help identify and mitigate threats but also protect consumers' privacy and data.

7.3 Mitigating Bias in AI Training Data for Cybersecurity Applications

Though AI is highly beneficial in transforming cybersecurity operations as we discussed in previous chapters, they are efficient only up to the quality and quantity of data they are being trained on. AI offers unmatched accuracy and speed in automated threat detection, response, and prevention tasks, but if the training data is biased, then it can lead to inefficient AI models that might not detect certain threats, flag genuine activities as malicious, or even unfairly target people from specific demography.

In this chapter, we will discuss more about biases in AI training data for cybersecurity applications and explore various ways to mitigate such biases.

7.3.1 Understanding Bias in AI Training Data

Definition

Bias in AI means errors in the AI model's output resulting from the already present prejudices in the training data. Such biases in the training data and AI models can lead to unfair, inaccurate, and misleading outcomes.

Types of Bias

Here is a quick overview of the different types of bias we can find in AI models:

1. Data Bias: It arises from the non-representative training data that is not complete and does not have proper details related to real-world scenarios.
2. Label Bias: It occurs when humans influence the labeling process through their prejudices and errors.

3. Algorithmic Bias: This kind of bias can arise due to the way algorithms process data to identify patterns. They often exaggerate the biases present in data.

4. Deployment Bias: This refers to biases that come out when AI systems are deployed in a different environment that do not represent the training data.

7.3.2 Sources of Bias in Cybersecurity Training Data

Historical Data

Historical data is often used to train the AI models for cybersecurity. There are chances that these data may contain inherent biases related to past security practices and threat intelligence. For example, if the historical data only focuses on specific types of threats and attack factors, the resulting AI model may be hinged toward such patterns and may ignore the newer and emerging types of threats.

Sampling Bias

When the data collected for training does not represent the actual environment the AI systems are deployed in, then sampling bias occurs.

For example, training data may put more emphasis on certain types of network traffic and user behavior and thus the model might perform inefficiently for other types of threat vectors.

Human Error in Labeling

Manual labeling of training data can also lead to bias, particularly in the inconsistent labeling process and labeling where it is influenced by the subjective judgments of the annotator. Such errors lead to mislabeled data that ultimately impacts the model's performance where it becomes less effective in identifying or classifying threats.

7.3.3 Impact of Bias on Cybersecurity AI Models

False Positives and Negatives

This is one of the major concerns related to biases in AI models. If the training data is biased, then the AI models can generate false positives or false negatives i.e., they may identify genuine activities as malicious and vice versa. This ultimately leads to inefficient use of resources, increases alert fatigue among cybersecurity professionals, and might even lead to security breaches.

Discriminatory Practices

Bias in AI training data may also lead to discriminatory practices i.e., it can target certain groups or behaviors as unfair. For example, AI systems can mark network traffic from a particular geographic region as malicious based on the training data used.

Erosion of Trust and Adoption

Biases present in AI models often lead to the erosion of trust among users and stakeholders. If security professionals find that AI models are ineffective and present errors, they perceive it as unreliable and might not readily use it for their security operations. Thus, it will lead to undermining the great benefits AI systems can offer in the field of cybersecurity.

7.3.4 Strategies to Mitigate Bias in AI Training Data

Data Collection and Preprocessing

- **Diverse Data Source**

Organizations and security professionals must try to collect training data from various sources that represent a variety of real-world threat scenarios, and behaviors. It will help create a more diversified and representative dataset.

- **Data Augmentation**

Security teams can use data augmentation techniques and artificially increase the diversity of training data. They can use Generative AI tools to generate synthetic data (as discussed in the introductory chapter) and augment existing data with variations.

- **Anonymization**

This means sensitive information must be removed from the training data so that models do not display biases towards specific groups or individuals.

Labeling and Annotation

- **Standardized Labelling Procedures**

Organizations should develop and enforce a standardized labeling procedure. This will help in minimizing human error and subjective judgments in the labeling process.

Crowdsourcing and Expert Reviews

It is also recommended that a combination of crowdsourcing and expert reviews should be used to label data. This will prove to be very effective in balancing out individual biases as well as improving the accuracy of labeled data.

Bias Detection in Labels

They must also implement techniques that can detect and correct biases in the labeled data. They can use statistical analysis to identify anomalies and inconsistencies in the labels.

Algorithmic Techniques

Detecting and correcting bias in Algorithms

Security teams must also incorporate algorithms that can detect and correct bias in training data and model outputs. Some popular techniques to mitigate bias in algorithms are re-weighing, re-sampling, and adversarial training.

Fairness-Aware Machine Learning Models

Organizations must aim to develop and deploy fairness-aware machine learning models that can minimize bias. Such models use fairness constraints and objectives during the training process.

Continuous Monitoring and Evaluation

Continuous monitoring and evaluation help to maintain the effectiveness and fairness of AI models in cybersecurity. Regular audits can be used to assess the AI model's performance and see if any kind of bias has emerged over time or not. Such audits may involve comparing the system's output to the expected performance standards and analyzing if they exhibit any kind of unintended discriminatory patterns.

Another way to ensure bias is not present is by gathering user feedback. The concerned teams must solicit input from end-users that will help them gain insights about the real-world application of AI systems, identify potential biases, and areas of improvement.

Over time, the system will autonomously learn and update its models to stay relevant and minimize bias.

Chapter 8

Ensuring AI Cybersecurity Accountability and Transparency

Whether it is now or in the future, the use of artificial intelligence will only grow be it the cybersecurity field or any other industry. In the world of cybersecurity particularly, it is embraced because AI makes operations faster and more efficient. But with the widespread use of AI, ensuring accountability and transparency is important to consider. As an AI or cybersecurity professional, everyone must be aware of the principles, have a clear understanding, and know how to establish mechanisms for accountability and transparency.

In this chapter, we will explore how AI makes decisions and addresses ethical concerns, strategies, and best practices to improve accountability and transparency that will strengthen the ethical foundations necessary for the responsible deployment of AI technologies.

8.1 Auditing AI Systems for Security Compliance

With many organizations now looking to deploy AI-powered cybersecurity systems for their defense against diverse cyberthreats, the need for a trusted audit mechanism has become crucial. Previously, we saw the complex interplay between AI and cybersecurity. Here, we will see how it necessitates proper vigilance to ensure it aligns with security compliance and standards.

It is very challenging when it comes to auditing AI systems for security compliance. Unlike traditional rule-based systems, AI systems are quite complicated, and dynamic and work in an opaque manner. This aspect of the AI system makes the assessment of its actions and decisions difficult. While the AI's capability to adapt and learn from data is the biggest asset of the AI system, they are the only reason that makes predicting and understanding its behavior complex. Therefore, ensuring AI-powered security systems comply with industry standards and regulatory frameworks becomes a difficult task.

One major challenge is understanding the complex AI algorithm that makes decisions related to cybersecurity. Especially in the case of certain machine learning models which are so opaque that it is very difficult to trace the decision-making process and identify potential vulnerabilities. Also, as the AI system is continuously evolving, it has become even more difficult to ensure these systems adhere to security standards.

8.1.1 AI Security Compliance: What is it?

AI Security Compliance refers to the process of aligning artificial intelligence (AI) applications in cybersecurity to the predefined regulations and standards that help to ensure their ethical and secure deployment. Over the time, when the use of AI in cybersecurity is deeply rooted, there is an absolute need for compliance frameworks to safeguard against potential risks and ethical concerns. This AI security compliance involves a proper examination of AI systems to verify; if they are adhering to industry-specific regulations, ethical guidelines, and predefined security protocols.

At its core, this compliance aims to maintain a balance between leveraging the innovative technology and efficiency offered by AI in cybersecurity and at the same time mitigating potential risks, protecting individual privacy, and ensuring ethical deployment of these systems.

8.1.2 Applicable Regulations and Frameworks

Many regulatory frameworks play an important role in shaping AI security compliance. As discussed in the previous chapter, the General Data Protection Regulation (GDPR) is an effective regulatory framework designed to protect users' data and privacy in the European Union from unethical usage, storage, and processing. This sets a difficult challenge for AI applications to navigate through. NIST Special Publication 800-161 provides guidelines for information security in the federal government, offering valuable insights into securing AI systems and networks.

Moreover, the International Organization for Standardization's ISO 27001 standard which focuses on information security management systems is relevant in the context of AI security compliance too. It provides a well-designed approach to managing sensitive information and aligns it with the imperative to protect data processed by AI algorithms.

It requires a great knowledge of AI technologies along with a commitment to ensure the principles of security, privacy, and ethical use are followed, to properly understand and implement these frameworks.

8.1.3 Challenges in Auditing AI for Security

- **Transparency and Explainability**

Certain AI models have inherent opacity that set forth the biggest challenge in auditing this AI system for security. This opaque nature is referred to as "black-box" models and they operate in a manner where their decision-making processes are not easy to understand or explain. The lack of transparency; poses a significant obstacle for auditors trying to examine the usefulness as well as ethical considerations of those AI applications. Innovative methods are needed to address such challenges to improve the transparency and explainability of AI algorithms which can ensure auditors can easily understand the decision-making process and perform effective security assessments.

- **Data Bias and Fairness**

This challenge arises right from the data set being used to train the AI systems. If the datasets are skewed and consist of biases, then it can introduce challenges in the auditing process. Biased data are responsible for discriminatory outcomes, negligence to comply with ethical regulations, and even exaggerating existing societal disparities, if any. So, auditors must check for any biases, identifying and rectifying them before they perform audits to ensure fair and equitable results. They must refine the algorithms to eliminate any kind of bias present, and then implement comprehensive strategies to select, pre-process, and validate the training data to align with ethical standards.

- **Cyberattacks and Vulnerability**

Cybera◎acks are another factor presenting a significant risk in auditing AI for security. Cybercriminals target vulnerabilities in AI systems and manipulate them to make incorrect decisions or reveal sensitive information. Therefore, auditors need to continuously identify and address potential vulnerabilities in real-time. Therefore, effective security measures including continuous monitoring, encryption, intrusion detection systems, etc. become important to protect the AI systems against various kinds of cyberattacks. So, the auditing process should not only focus on the functionality of AI models but also their resilience against these cyberthreats.

8.1.4 Approaches to Auditing AI for Security

Static and Dynamic Analysis

Static and dynamic analysis are needed to completely audit the AI system for security. While static analysis involves a proper examination of the AI model's code and architecture, dynamic analysis refers to the evaluation of the AI system's behavior in real-time. Static analysis facilitates code review, vulnerability scanning, and data input testing, helping to find out the potential vulnerabilities and weaknesses during the development phase. Whereas in dynamic analysis, adversarial simulations, penetration testing, and other real-world scenarios are simulated to evaluate the system's response. By incorporating this dual approach, auditors ensure a thorough scrutiny of both the inherent structure and real-world functionality of AI models.

Risk Assessment and Mitigation Strategies

When we want a robust auditing process, then we must ensure there is a systematic risk assessment framework that is customized to AI implementation. This process refers to the identification of potential threats and vulnerabilities that are specific to AI systems. By using existing frameworks like the NIST Cybersecurity Framework, auditors can easily categorize, assess, and prioritize risks. Once the risks are identified, proper mitigation strategies can be designed and implemented. It may include refining algorithms, improving data integrity, and strengthening access controls. A combination of risk assessment and auditing processes can help organizations proactively manage potential security challenges associated with AI deployment.

8.2 Explainable AI for Security Decision-Making

The opaque nature of AI decision-making also referred to as the Black Box model in the world of cybersecurity o◎en raises concerns regarding the trust and accountability of using this system. The complex AI algorithms that help in security decision-making must be easily interpretable otherwise they can introduce huge challenges in the transparency needed to gain user trust and ensure system accountability.

Explainable AI, therefore, becomes important in this context because it serves as the tool the enhance trust, refine accuracy of decision-making, and ensure compliance. By explaining the workings behind AI-driven security decisions, the explainable AI proves to be an important element that strengthens the ethical and effective deployment of AI technologies in the cybersecurity world.

8.2.1 Benefits of Explainable AI for Security Decisions

Enhanced Situational Awareness

- ***Deeper Threat Insights:*** Explainable AI doesn't simply flag an anomaly, but it tells you why it is suspicious. It offers details on specific Indicators of Compromise (IOCs) and a◎ack pa◎erns which helps security analysts quickly understand the nature of the threat and prioritize their response

- ***Uncovered Blind Spots:*** Explainable AI can identify hidden vulnerabilities and emerging threats that can go unnoticed by human analysts as these systems can analyze data beyond traditional security metrics. This helps in the early detection of potential threats and timely response to any kind of security breaches.

- *Predictive Capabilities:* Explainable AI helps to identify potential a◎ack vectors and predict future threats as it can learn from past incidents and leverage historical data for analysis. This predictive power helps the security teams to proactively strengthen security measures and allocate resources based on expected risks.

Improved Incident Response

- *Faster Resource Allocation:* Understanding the context and reasoning behind why AI-triggered alerts will help professionals perform more targeted and efficient responses. So, instead of wasting time on false positives, the cybersecurity teams can immediately focus on critical threats, and give prompt responses to mitigation efforts.
- *Optimized Investigation Workflows:* Knowing the reason behind why AI-triggered alerts can help analysts customize their investigation process which can save valuable time and resources. The explainable AI also helps in collaboration between humans and AI further strengthening effective incident response.
- *Adaptive Learning and Response:* Explainable AI continuously learns from new threats and adjusts its decision-making parameters which enables professionals to handle incidents and responses more dynamically.

Reduced False Positives and Negatives

- *Minimized Alert Fatigue:* As explainable AI can easily filter out noise and focus on genuine threats, it can significantly reduce the number of false alerts. This can free up analysts' time and resources from wasting on unnecessary events and they can concentrate on important security concerns.

- ***Precision Targeting:*** Explainable AI can highlight the specific data points and anomalies that triggered the alert which will help analysts understand the actual nature of the threat. It will allow them to differentiate between genuine and irrelevant events reducing the risk of mistakenly ignoring important threats due to false positives.
- ***Data-Driven Prioritization:*** Explainable AI can also prioritize events based on the severity of threats and their potential impact which will ensure critical issues are addressed on priority. Thus, this data-driven approach also helps to optimize resource allocation and prevent neglecting serious threats.

Building Trust with Stakeholders

- ***Enhanced Transparency and accountability:*** Explainability helps to build trust by helping decision-makers and non-technical stakeholders gain clear insights into the reasoning behind AI-driven security decisions. This transparency further enhances collaboration and strengthens buy-in for AI-powered security solutions.
- ***Reduced Bias and Discrimination:*** It also helps identify and mitigate potential biases within AI models that ensure security decisions made are fair and not influenced by external or irrelevant factors. This builds trust and promotes ethical security practices.
- ***Improved Communication and collaboration:*** Security teams can effectively communicate threat assessment and risk mitigation strategies to non-technical stakeholders if they clearly understand the workings of AI models and algorithms. This will help in strengthening the organization's overall security posture and effectively manage threats.

By providing deeper insights and building trust, explainable AI helps the security teams to make more informed decisions so that they can optimize their response workflows and ultimately improve overall security efforts.

8.2.2 Challenges of Implementing Explainable AI in Security

- **Complexity of AI models**

Many powerful AI models used in cybersecurity, like deep learning algorithms, operate as "black boxes." Their internal workings are opaque which makes it difficult to understand how they arrive at their conclusions. This lack of transparency affects the debugging process, identifying biases, and ultimately trusting their recommendations.

Even with explainable models, analyzing the contribution of specific features to the final decision can be a difficult task. Identifying the key factors that trigger alerts or influence assessment is important for effective security action. However, the interpretation of complex features can be challenging for human analysts.

- **The trade-off between accuracy and Explainability**

It is difficult for the simple models to achieve the same level of accuracy as the complex and black-box models can achieve. This creates a trade-off where if you prioritize explainability then the effectiveness of threat detection and incident response could potentially be compromised.

To make explanations more easily comprehendible, developers may try to oversimplify the reasoning behind the AI's decisions leading to incomplete explanations that cannot explain the full complexity of the decision-making process.

Therefore, developing and deploying explainable AI models require additional computation resources and development time. This poses a significant obstacle for organizations that are already limited with their resources in the cybersecurity infrastructure.

Data Privacy and Security Considerations

When you try to explain the reasoning behind AI systems, you might be involved in revealing some of the data used to train the AI models for decision-making. So, if the data contains sensitive information like personally identifiable information (PII), or confidential business data, then it will become problematic. Balancing explainability with data privacy requires careful consideration and implementation of secure data anonymization techniques.

There is a risk of creating new vulnerabilities as explainability can expose the inner workings of the AI model. Cyberattackers can exploit this knowledge to design new targeted attacks and manipulate the model's decision-making process which can pose significant security risks.

Even if they do not directly exploit the explanation mechanism, the cyberattackers can identify patterns in the explanations themselves. This will reveal valuable information about the data or vulnerabilities in the AI system. So, it becomes very important to implement a robust security measure alongside explainability implementations.

These challenges highlight how complex the implementation of explainable AI is in the field of cybersecurity. So, it is important to find the right balance between transparency, accuracy, and security by the developers and security professionals alike.

8.2.3 Approaches to Explainable AI for Security Decisions

Interpretable Models and Algorithms

- ***Rule-based Systems:*** These systems rely on clear, human-readable rules to make decisions. In the context of security, this integrated transparency in the ruled-based system allows one to easily understand and grasp the logic behind why an alert and how the risk assessment is done. But, as mentioned previously, their capabilities are limited as compared to more complex models.
- ***Decision Trees:*** Just like rule-based systems, decision trees also offer similar visual representations of their decision-making process and branching based on specific conditions. This helps to understand the factors that influenced AI's recommendation and why it did not take the potential paths. These systems are quite powerful for specific tasks but can lag for complex decisions.
- ***Linear Models:*** In Linear Models, the outputs are mapped to inputs using readily interpretable linear equations. Because of this, it becomes easier to understand the relative influence of different features on why the decision is taken, thus helping to prioritize security actions. These models can also have limited effectiveness in complex real-world security scenarios.

Visualization and Feature Attribution

- ***Heatmaps and Saliency Maps:*** These visual techniques are used to highlight the areas of inputs such as an image or network traffic which could have the most impact on AI's decision. Through these maps, the analysts can locate the exact position of the specific data points that trigger alerts or influence the risk assessment process.

- ***Feature weights and contribution charts:*** This technique helps to quantify the relative influence of different features on the final decision. So, analysts get a be◎er understanding of how different data points contribute to the overall assessment enabling them to focus on the most important factors.
- ***Interactive Dashboards:*** If these visualization techniques and interactive interfaces are combined, it can create a highly powerful dashboard for analysts that can help them explore AI's reasoning in depth, and adjust parameters and filters to gain different perspectives on the decision-making process.

Counterfactual Explanations and Scenario Analysis

- ***What-if Scenarios:*** These techniques are employed to understand how AI's decision would change if a few input parameters would differ. Therefore, it helps to gain insights into the underlying logic and evaluate the potential impact of protective actions.
- ***Contrastive Explanations:*** With this technique, the features of suspicious events are compared to those of normal events. This highlights the key differences that triggered alerts and helps identify anomalies and understand the specific pa◎erns that AI has learned to associate with threats.
- ***A▯ack-Resistance Explainable AI:*** This involves developing AI models and techniques that cannot be manipulated by a◎ackers. It will ensure that explanations remain reliable even when dealing with compromised data or analyzing a◎acks that were aimed to exploit the explainability system.

By employing these different explainable AI approaches, organizations can enhance the potential of explainable AI for making informed security decisions. When interpretable models are combined with visual techniques and scenario analysis, then it can do wonders by offering a multi-faceted understanding of AI's reasoning. Ultimately, it will enable cybersecurity professionals to build trust, optimize response strategies, and strengthen their overall security posture.

8.3 Developing Explainable AI for Improved Security Decision-Making

In the previous chapters, we saw the importance of Explainable AI. One of the major problems of implementing AI models and solely relying on them is the lack of explainability in their decision-making process. The users and stakeholders do not know how the AI models have concluded. And therefore, the need for developing Explainable AI is larger than ever. It aims to bridge the gap between human intuition and algorithmic black boxes.

In this chapter let us explore the importance of XAI in cybersecurity in detail and understand various techniques for developing a more transparent and understandable AI model.

8.3.1 Why explainability matters in Cybersecurity?

Traditional AI models use complex algorithms that process data and arrive at conclusions, but they do not provide a clear explanation for their decisions or reasoning. This limitation of older models introduces several challenges in cybersecurity as follows:

- **Limited Trust and Adoption**

Security teams might not readily accept AI-powered security solutions if they are not convinced about how the model arrives at a specific threat detection or response decision. If there is no clarity then it might hinder widescale adoption of AI systems for cybersecurity as well as reduce their effectiveness in real-world security operations.

- **Debugging and Improvement**

Whenever the AI system makes a mistake such as a false positive or false negative, it becomes even more necessary to understand why it is making such mistakes and the root cause for effective debugging and improvement. If there is no proper explainability, then troubleshooting becomes a highly complex and time-consuming process.

- **Accountability and Compliance**

Some of the security decisions can pose significant consequences. In a few cases, regulations might require organizations to explain the reason behind taking security actions. In such cases, unexplainable AI models can make it very challenging to demonstrate compliance and accountability.

8.3.2 Techniques for Developing Explainable AI Models

As technology is ge◎ing advanced, the need as well as techniques for XAI is also constantly evolving. Here are some of the popular techniques used to make AI models more explainable:

- **Feature Importance**

In this technique, they highlight certain specific features or data points that influenced the AI models the most in their decision-making process. So, security analysts can easily understand which factors contributed the most to the model's classification, be it malicious or genuine.

- **Rule-based Explanations**

There are many cases where AI models can be converted into a set of rules. This set of rules further represents the logic behind the model's decision-making process. Though it is not feasible for all AI models, it can help explain specific decisions that are easy to understand by humans.

- **Counterfactual Explanations**

In this approach, the focus is on how the hypothetical change in the input data would have affected the output. The security teams then analyze these "what if" scenarios and gain the necessary insights about the model's reasoning.

8.3.3 Human-Centered Approach to Explainable AI in Cybersecurity

Developing XAI for cybersecurity needs a human-centered approach and must consider the demands of security analysts. For example, the explanations should be customized to as per the user's demand and contain the details and format of explanations as needed by the user/audience. While the security analysts might need more technical explanations, the decision-makers and non-technical stakeholders might be happy with high-level summaries.

On top of that, there should be interactive visualizations that can present even the most complex data and other explainable metrics in an easy-to-understand and digestible format. It will help to properly understand the AI model's reasoning according to human understanding levels.

Another important thing is the involvement of humans. No ma◎er how advanced the explainable AI (XAI) becomes, the need for human interventions will always be important. Security analysts must use various explainability tools to understand the output presented by AI models and then only make informed decisions by employing their knowledge and expertise in their domain.

8.3.4 Development of XAI Beyond Technical Advancements

To develop and reap the full benefits of explainable AI, it is necessary to go beyond just the technical advancements. There is a need for a cultural shi◎ within the cybersecurity community. Organizations must try to foster an environment of collaboration between AI engineers or researchers and security professionals. This will ensure that the XAI solutions being developed meet the specific requirements and challenges in the cybersecurity industry.

Apart from that, education is another essential thing to consider. Every cybersecurity professional must be trained in the various Explainable AI concepts and techniques that will help them properly use this incredible tool and understand the insights it provides. They should be made aware of the limitations and capabilities of such models. This transparency helps to build trust in AI systems. Most importantly, there should be open communication between security teams and stakeholders regarding AI-powered security decisions.

Explainable AI offers a transformative opportunity for the future of the cybersecurity industry. Organizations must try to develop and incorporate XAI and must encourage collaboration between AI and cybersecurity teams to introduce innovative solutions.

CHAPTER 9

Shielding Against AI: Defenses in the Cyber Arms Race

Introduction

As we turn the page from AI ethics, a cornerstone of our previous discussions, we find ourselves standing on the cusp of an advanced era where AI-centric cybersecurity strategies are beneficial and essential. The genesis of AI has heralded a sea change in the cybersecurity landscape; a domain once navigated by experts behind screens now stands augmented by an invisible yet potent AI-driven force.

Imagine playing a game of digital chess where a shadowy opponent mirrors every move you make, anticipating your strategy even before you lay a finger on a piece. This notion is the reality of modern cybersecurity, where organizations wield AI as a shield, fortifying their digital realms against a barrage of cyberattacks that grow more intricate by the hour. Meanwhile, adversaries, with equal cunning, harness AI to craft attacks of unprecedented sophistication.

To illustrate this point, consider a hypothetical incident involving a significant financial institution. This organization, equipped with the latest security technology, encountered a deceptively benign code. Beneath its harmless appearance, however, lay a learning algorithm with a sinister purpose: to observe quietly, gather intelligence, and prepare to launch a precision strike on the institution's digital assets. This AI-enhanced cyberattack was designed to evade detection, silently slipping past traditional security measures, and revealing the vulnerability even in the most fortified systems.

This chapter is our expedition into this unfolding narrative, the cybersecurity arms race, a relentless push and pull that marks the rhythm of this digital era. As we dissect this landscape, we will encounter the yin-yang of AI-enhanced cyberattacks and AI-augmented cyber defense operations and grapple with the ethical quandaries they pose.

We will pivot through the digital battlegrounds, where AI not only strengthens the ramparts against incoming assaults but also serves as the tip of the spear for offensive cyber operations. By examining these emerging threats and the defense mechanisms designed to thwart them, we will peel back the layers of AI's impact and chart the trajectory of its role in the theater of digital warfare.

The stakes are high, and the adversaries are evolving. AI offers a beacon of hope, illuminating the vulnerabilities we must seal and the threats we must anticipate. As we marshal our forces in this chapter, let us prepare to delve into the heart of the fray, understanding that in the arms race of cybersecurity, innovation is our most formidable ally. By examining emerging threats and cutting-edge defense mechanisms, we will uncover the intricacies of AI's role in shaping the future of digital warfare.

9.1 Defending Against AI-Enhanced CyberAttacks

Chapter 9 examines the proactive measures cybersecurity professionals take to counter the evolving landscape of AI-enhanced cyber threats. It highlights the latest developments leveraging sophisticated AI-driven defenses to shield critical systems and data from increasingly intelligent and adaptive attacks. The chapter explores how experts deploy cutting-edge technologies and strategies in this continuous arms race between cyber defenders and attackers.

As we navigate through layers of proactive defenses and reactive measures, we gain insight into the complex interplay of innovation and resilience that defines modern cybersecurity efforts. This exploration fortifies our digital strongholds and evolves our defenses to counter the threats they are designed to thwart.

9.1.1. The Evolution of AI in Cybersecurity

AI technologies are undergoing rapid evolution, fueling a burgeoning arms race within the cybersecurity landscape (McAfee, 2024). This escalating confrontation plays out on two fronts. Malicious actors, on one side, leverage AI to augment their cyberthreats, rendering them progressively more sophisticated, elusive, and intricate to detect (Wang et al., 2023). Concurrently, cybersecurity professionals wield the power of AI to fortify cyber defenses, bolstering threat detection, response, and prevention capabilities.

On the offensive front, cybercriminals relentlessly explore novel strategies to weaponize AI for nefarious purposes. They develop AI-powered malware that can mutate and adapt, effortlessly evading traditional security measures. These malicious AI algorithms enable automated reconnaissance, meticulously identifying vulnerabilities within systems and networks and cra◎ing tailored exploits to target those weaknesses with surgical precision.

Moreover, AI empowers cybercriminals to generate compelling social engineering a◎acks, such as meticulously personalized phishing emails and deepfake videos, designed to deceive even the most vigilant individuals. Leveraging sophisticated language models and generative adversarial networks (GANs), these AI-driven a◎acks create compelling and realistic content, significantly increasing the likelihood of successful breaches.

Distributed denial-of-service (DDoS) a◎acks also benefit from the integration of AI, enabling the coordinated orchestration of vast botnets and the intelligent variation of a◎ack vectors. This adaptability poses formidable challenges for traditional DDoS mitigation techniques, as the a◎acks can quickly dynamically shi◎ and evolve, overwhelming defense mechanisms.

Furthermore, AI techniques like machine learning aid cybercriminals in breaking encryption and cracking passwords, posing grave risks to data confidentiality and system integrity. These AI-powered a◎acks can rapidly cycle through vast combinations of keys and passwords, exploiting even the most complex encryption algorithms with remarkable efficiency.

On the defensive front, cybersecurity professionals harness the power of AI to reinforce their cyber defenses. AI algorithms are pivotal in developing advanced intrusion detection and prevention systems (IDS/IPS). These systems leverage machine learning to analyze network traffic pa◎erns, identify anomalies, and detect sophisticated real-time threats, enabling proactive response and mitigation efforts.

AI's processing of vast amounts of security data enables efficient threat hunting and forensic analysis. Cybersecurity teams can leverage AI algorithms to correlate disparate security events, uncover hidden pa◎erns, and derive actionable insights for incident response and remediation, enhancing their ability to thwart emerging threats.

Additionally, AI has revolutionized vulnerability management by automating the identification and prioritization of software vulnerabilities. AI-powered security solutions can proactively scan for weaknesses, assess their severity, and recommend remediation strategies, streamlining patch management processes and reducing the risk of exploitation.

As AI capabilities evolve rapidly, cybersecurity professionals actively explore new frontiers, such as AI-driven deception techniques, automated threat hunting, and predictive security analytics. These innovations strive to stay ahead of the curve by anticipating and mitigating emerging threats before they significantly damage systems and networks.

The ongoing arms race between offensive and defensive applications of AI in cybersecurity highlights the critical importance of responsibly developing and deploying these technologies. It is crucial to strike the right balance between harnessing AI's immense potential and mitigating its associated risks to maintain a robust and resilient cybersecurity posture in an increasingly complex and constantly evolving threat landscape.

9.1.2 The Arms Race in AI-Enhanced CyberThreats

Cybercriminals continuously seek to weaponize advanced artificial intelligence (AI) capabilities, pushing the boundaries of cybersecurity threats. A significant concern is the rise of AI-powered malware that can mutate and adapt, evading traditional security measures. These malicious algorithms analyze defenses and evolve tactics, exploiting vulnerabilities before teams respond effectively.

AI algorithms also enable large-scale automated reconnaissance, rapidly scanning networks and systems for potential attack vectors like software flaws and misconfigurations. With this intelligence, cybercriminals craft highly tailored exploits to maximize breach success.

AI language models enhance social engineering attacks by generating convincing, personalized phishing emails that mimic legitimate communications. Deepfake technology powered by AI can create realistic videos impersonating individuals for social engineering or disinformation campaigns.

AI coordinates massive botnets for distributed denial-of-service (DDoS) attacks, varying tactics to overwhelm mitigation systems. AI password cracking leverages pattern recognition to rapidly cycle through encryption keys, compromising data confidentiality.

As AI capabilities grow, so does the potential for cybersecurity misuse. Cybercriminals continually explore novel AI weaponization, outpacing defensive measures. This evolving landscape demands proactive, adaptive cybersecurity, where defensive AI techniques counter emerging threats.

Responsible AI development, robust governance frameworks, and international cooperation are crucial to mitigate AI-powered cyber risks. Balancing AI's potential benefits while addressing risks is vital for a secure cyberspace.

9.1.3 Historical Perspectives on the Use of AI for Cyber Defense

While the potential misuse of AI in cybersecurity raises concerns, the technology has also been instrumental in enhancing cyber defenses. Historically, AI has been crucial in developing advanced intrusion detection and prevention systems (IDS/IPS). These systems harness machine learning algorithms to scrutinize network traffic patterns, pinpoint anomalies, and uncover sophisticated threats as they happen.

An AI algorithm processes vast amounts of security data, enabling efficient threat hunting and forensic analysis. They can correlate disparate security events, uncover hidden patterns, and provide actionable insights for incident response teams.

Furthermore, AI has revolutionized vulnerability management by automating the identification and prioritization of software vulnerabilities. AI-powered security solutions can proactively scan for weaknesses, assess their severity, and recommend remediation strategies, streamlining patch management processes.

As AI capabilities advance, cybersecurity professionals explore new frontiers, such as AI-driven deception techniques, automated threat hunting, and predictive security analytics. The ongoing evolution of AI in cybersecurity highlights the need for responsible development and deployment of these technologies to maintain a robust and resilient cybersecurity posture.

9.2 Understanding AI-Enhanced Threats

As the adoption of artificial intelligence (AI) technologies continues to accelerate, cybercriminals actively explore ways to leverage these powerful capabilities for malicious purposes. Cybersecurity professionals and organizations must develop a comprehensive understanding of AI-enhanced threats to defend against them effectively.

9.2.1 Characteristics of AI-Powered CyberThreats

Understanding AI-enhanced cyberthreats requires dissecting their unique characteristics. These threats go beyond simple automation; as they leverage AI's capabilities to become more intelligent and adaptive. Here is a breakdown of what makes them so dangerous:

1. **Evolving Tactics:** AI can analyze vast amounts of data on security measures, allowing attackers to refine their tactics constantly. These threats can morph and adapt to bypass traditional defenses, making them highly dynamic and unpredictable.

2. **Social Engineering on Steroids:** AI can personalize phishing attacks with unsettling accuracy. By analyzing social media profiles and communication patterns, attackers can craft highly believable emails or messages that bypass human suspicion.

3. **Weaponized Automation:** AI can automate complex attack sequences, allowing attackers to launch large-scale assaults with minimal effort. This automation can overwhelm traditional security defenses and significantly increase the impact of an attack.

9.2.2 Real-World Examples: AI in Action

To truly grasp the threat, let us explore some real-world case studies:

1. **Deepfakes for Disinformation:** In 2020, attackers used AI-generated deepfakes to impersonate a CEO in a fraudulent voice call, successfully tricking an employee into transferring a large sum of money.

2. **AI-powered Ransomware:** Recent ransomware variants have incorporated AI to identify and target high-value assets within a network, significantly increasing the potential damage caused by an attack.

These are just a few examples, but they highlight the growing sophistication of AI-powered threats. As AI continues to evolve, so will attackers' methods. Staying vigilant and continuously adapting our defenses is crucial in this ever-changing landscape.

9.3 Defense Strategies Against AI Attacks

9.3.1 AI-Driven Threat Modeling and Risk Assessment

Security professionals proactively leverage AI to model potential a◉ack vectors and assess associated risks. AI algorithms process extensive data, incorporating threat intelligence, vulnerability details, and system configurations to pinpoint potential weaknesses and a◉ack vectors. These AI-driven models simulate various a◉ack scenarios, enabling security teams to prioritize mitigation efforts and allocate resources effectively.

Furthermore, AI assists in continuous risk assessment, dynamically updating threat models as new vulnerabilities emerge or a◉ack pa◉erns evolve. This proactive approach empowers organizations to avoid emerging threats and implement preventive measures before a◉acks occur.

9.3.2 Building Resilient AI Systems for Cyber Defense

To counter the threat of adversarial AI a◉acks, security teams develop resilient AI systems designed explicitly for cyber defense. These systems incorporate robust training techniques, such as adversarial training, which exposes AI models to carefully cra◉ed adversarial examples during training. This approach enhances the models' ability to detect and resist adversarial a◉acks, making them more robust and reliable.

Additionally, security professionals implement AI systems with built-in safeguards and security controls. These measures include secure data handling protocols, access control mechanisms, and rigorous testing and validation processes. By integrating security from the ground up, AI systems for cyber defense can operate with high integrity and reliability, minimizing the risk of compromise or manipulation.

9.4 Detection and Mitigation Techniques

9.4.1 Utilizing AI for Anomaly Detection and Response

Cybersecurity professionals leverage the power of AI to enhance anomaly detection and response capabilities, providing a robust defense against emerging cyberthreats, including those powered by AI.

AI algorithms process extensive data, recognizing patterns and flagging deviations from established norms. Security teams harness these capabilities by deploying AI-driven anomaly detection systems to continuously monitor network traffic, system logs, and user behavior patterns.

These AI systems employ advanced machine learning techniques, such as unsupervised learning algorithms, to establish baselines of regular activity within an organization's networks and systems. The AI models can identify anomalies indicating potential threats or compromises by understanding typical behavior.

Upon detecting an anomaly, AI-powered security systems can initiate automated response actions tailored to the threat. These actions may include:

1. **Traffic Quarantine**: Suspicious network traffic can be isolated and quarantined, preventing potential malware or unauthorized access from spreading further within the network.
2. **System Isolation**: Compromised systems or endpoints exhibiting abnormal behavior can be automatically isolated from the network, containing the threat, and preventing lateral movement.

3. **Incident Response Procedures**: AI systems can trigger predefined incident response procedures, alert security teams, and initiate investigation and remediation efforts.
4. **Adaptive Countermeasures**: AI models can analyze the detected threat and dynamically deploy countermeasures, such as firewall rule updates, endpoint protection policy changes, or security patch deployments, to mitigate the identified risk.

Moreover, AI-driven anomaly detection systems continuously learn and adapt from observed attacks and security incidents. By analyzing successful and failed attack attempts, these systems can refine their detection models, improve threat signatures, and enhance their ability to identify and mitigate similar threats in the future.

This adaptive learning process ensures that defensive AI systems remain effective against evolving cyberthreats, reducing the likelihood of successful attacks over time. Furthermore, these systems can integrate feedback from human security analysts, incorporating domain knowledge and expertise into their learning process.

By leveraging AI for anomaly detection and response, organizations can identify and mitigate cyberthreats, including those powered by AI, in a timely and effective manner. This collaborative approach, combining the strengths of AI and human expertise, creates a formidable defense against the ever-evolving landscape of cyberthreats.

Adaptive AI Systems that Learn from Attacks to Bolster Defenses

As cyberthreats keep changing and getting more advanced, even using AI, security teams are using flexible AI systems. These AI systems can learn from the attacks they see and keep improving at protecting against threats over time.

These adaptive AI systems leverage advanced machine learning techniques to analyze successful and failed a◎ack a◎empts, extracting valuable insights and pa◎erns from adversaries' tactics, methods, and procedures (TTPs). By studying the intricacies of these a◎acks, AI models can identify new a◎ack vectors, refine threat signatures, and enhance their ability to detect and mitigate similar threats in the future.

As new a◎acks are observed and analyzed, the adaptive AI systems update their detection models and defensive strategies accordingly. This continuous learning and adaptation process ensures that the defensive AI capabilities remain effective against evolving cyberthreats, reducing the likelihood of successful breaches over time.

Moreover, these adaptive AI systems integrate feedback and guidance from human security analysts and incident response teams. Security professionals can provide domain knowledge, expert insights, and contextual information about specific a◎acks or threat actors. The AI systems then incorporate this valuable input into their learning process, combining the strengths of human expertise and machine learning capabilities.

When AI systems and human analysts work together, they form a cooperative partnership that benefits both sides. The AI systems and human experts work together to improve each other. AI learns from security professionals' knowledge while giving them helpful information and suggestions for be◎er decisions. This concept is hugely important because some think leveraging AI systems is cheating. Society is moving towards a point where humans work with AI to increase continuous learning and produce a be◎er product.

Furthermore, adaptive AI systems can leverage techniques such as adversarial training, where trainers expose the models to carefully cra◎ed examples during training. This approach enhances the AI systems' robustness and resilience, improving their ability to detect and resist adversarial a◎acks, including those powered by AI.

Adaptive AI systems can proactively identify and mitigate emerging cyberthreats by continuously learning from observed attacks and integrating human expertise. As new attack vectors and techniques emerge, these AI systems can rapidly adapt and deploy countermeasures, minimizing the potential impact of successful breaches. The cyclic nature of AI, rooted in machine learning, endows it with its power; it continuously learns and adapts to its surrounding environment.

Implementing adaptive AI systems is a crucial component of a comprehensive cybersecurity strategy. This strategy allows organizations to have strong defenses against all types of cyberthreats, even the ones that use advanced AI techniques. As threats keep changing, these systems help organizations protect themselves.

9.5 Developing a Proactive Posture

In the face of an evolving cybersecurity landscape, where threats are becoming increasingly sophisticated and AI-enhanced, organizations must adopt a proactive defense posture to stay ahead of potential attacks. This proactive approach leverages the power of artificial intelligence (AI) to anticipate and mitigate threats before they can cause significant harm.

Incorporating AI Into Proactive Cybersecurity Strategies

Security teams integrate AI technologies into their proactive cybersecurity strategies, leveraging the capabilities of these advanced systems to enhance threat intelligence, vulnerability management, and risk assessment processes.

AI programs study huge amounts of data, like threat reports, lists of software vulnerabilities, and security logs. This process allows AI to spot possible ways hackers could attack and identify new threats as they emerge. This intelligence enables security teams to prioritize mitigation efforts and implement preventive measures before adversaries can exploit vulnerabilities.

AI-driven risk assessment models simulate attack scenarios tailored to an organization's systems and vulnerabilities. These simulations provide insights into the potential impact of threats, helping security teams effectively allocate resources and implement defensive measures.

Role of AI In Developing Anticipatory Defense Mechanisms

AI plays a crucial role in developing anticipatory defense mechanisms that can proactively detect and mitigate cyberthreats, including AI-powered ones.

Security teams employ AI systems to continuously monitor network traffic, system logs, and user behavior patterns, establishing baselines of regular activity. These AI models leverage advanced machine learning techniques, such as anomaly detection algorithms, to identify deviations from established norms that may indicate potential threats or compromises.

Upon detecting anomalous activity, AI systems can initiate automated response actions, such as quarantining suspicious traffic, isolating compromised systems, or triggering incident response procedures. This rapid and automated response capability enables proactive threat containment and mitigation, minimizing the potential impact of successful attacks.

Furthermore, security teams employ AI algorithms to develop proactive defense mechanisms against AI-powered cyberthreats. They utilize techniques such as adversarial training, where AI models undergo training by exposing them to carefully crafted adversarial examples. This adversarial training approach fortifies the robustness and resilience of the

AI systems, bolstering their capabilities to detect and resist AI-enhanced attacks effectively.

Additionally, AI systems continuously learn and adapt from observed attacks, refining their detection models and defensive strategies. This adaptive learning process ensures that the anticipatory defense mechanisms remain effective against evolving cyberthreats, including those leveraging AI capabilities.

By incorporating AI into proactive cybersecurity strategies and developing anticipatory defense mechanisms, organizations can stay ahead of potential threats, including AI-powered ones. By acting early, security teams can find and reduce risks before they cause significant problems. This method gives them tough and adaptable cybersecurity defenses, even as threats change.

9.6 Ethical Considerations in AI Defense

As more companies start using AI for cybersecurity, it's essential to carefully consider the ethical issues when using such powerful AI technologies for defense and protection. Organizations must strike a delicate balance between implementing aggressive AI-driven defense mechanisms and adhering to ethical constraints and principles.

Balancing Aggressive AI Defense with Ethical Constraints

Security teams deploy AI systems to proactively detect and mitigate cyberthreats, including those enhanced by AI capabilities. However, these defensive AI algorithms must operate within well-defined ethical boundaries to protect individual privacy, uphold civil liberties, and maintain public trust.

Robust governance frameworks and ethical guidelines establish clear principles for developing and deploying AI defense systems. These guidelines outline data collection, handling, and usage standards, ensuring that the AI models do not inadvertently infringe on personal privacy or exploit sensitive information unethically.

Moreover, security professionals must carefully consider the potential consequences of aggressive AI defense mechanisms. Security teams must apply automated response actions judiciously, such as isolating compromised systems or quarantining suspicious traffic, to mitigate threats effectively while avoiding unintended disruptions or collateral damage.

Rigorous testing, validation, and oversight processes ensure that AI defense systems operate as intended, minimizing the risk of undesirable outcomes or unintended harm. Additionally, these systems incorporate human oversight and control measures, allowing security analysts to intervene and make informed decisions when necessary.

AI and Privacy: Defending without Overstepping

Integrating AI technologies in cybersecurity raises concerns regarding privacy and the potential misuse of personal data. Security teams must implement robust privacy-preserving measures to ensure that their AI defense systems operate transparently and respect individual privacy rights.

Unique methods allow AI systems to learn and make predictions without directly seeing or storing people's private and sensitive information. These methods include differential privacy, secure multi-party computation, and federated learning. The key is that AI never gets full access to personal data, which protects people's privacy. These approaches allow for effective threat detection and mitigation while minimizing the risk of privacy violations.

Furthermore, organizations must establish clear data governance policies and implement strict access controls to protect the privacy of individuals whose data AI defense systems may process. Regular inspections confirm the organization obeys all privacy laws and regulations and follows the right processes.

Transparency and open communication with stakeholders, including employees, customers, and the public, foster trust and understanding regarding the use of AI in cybersecurity defense. Organizations need to get ahead of people's worries about AI. They should teach team members and the public about it. They must also show that the organization is commi◎ed to using AI responsibly and ethically.

By carefully balancing aggressive AI defense strategies with ethical constraints and privacy considerations, organizations can harness AI's power to bolster their cybersecurity posture while maintaining public trust and upholding fundamental rights and values.

9.7 Case Study: AI Defense in Action

One notable case study highlighting the effectiveness of AI defense in action is the disruption of the Emotet botnet in early 2021. Emotet was a highly sophisticated and elusive malware strain that had plagued organizations worldwide for years, causing significant financial losses and data breaches.

Security researchers leveraged advanced AI algorithms to analyze the complex network of compromised systems and identify intricate pa◎erns in Emotet's operations. This AI-driven intelligence enabled law enforcement agencies and cybersecurity firms to collaborate and execute a coordinated global operation, effectively dismantling the botnet's infrastructure and disrupting its command-and-control servers.

The successful takedown of Emotet demonstrated the power of AI in thwarting sophisticated cyberthreats. AI algorithms played a crucial role in unraveling the malware's complex behaviors, tracking its communications, and ultimately neutralizing a formidable threat that had evaded traditional defense measures for an extended period.

CHAPTER 10

The Edge of Offense: AI-Augmented Cyber Operations

Introduction

Cyberspace is undergoing a revolution with the integration of AI into offensive capabilities. This revolution raises critical questions. Ethically, AI attacks can cause widespread civilian harm, blurring lines between military targets and innocent bystanders. International law must address AI's role in cyberwar, making attribution and accountability unclear.

Strategically, the ease of using AI for attacks could lower the threshold for conflict as nations scramble to maintain a technological edge, potentially fueling instability in the international order (Slaughter et al., 2020). To illustrate this growing threat, we will explore specific tools like deepfakes (i.e., manipulated videos for disinformation) (Shao et al., 2023), AI phishing (i.e., highly targeted email scams) (Li et al., 2022), and even autonomous attack drones capable of independent network disruption. By examining these tools and the broader strategic landscape, we can gain a comprehensive understanding of the complexities of AI-powered cyber warfare.

10.1 The Offensive Use of AI: Overview and Implications

The rapid advancement of AI has opened new frontiers in cyber warfare. As nations and non-state actors increasingly integrate AI into their offensive capabilities, it is crucial to examine the potential implications and challenges arising from this technological shi◎.

Ethical, Legal, and Strategic Dimensions of AI in Offensive Operations

Integrating AI in offensive military operations raises many ethical, legal, and strategic considerations. As AI systems grow more advanced and autonomous, using them in offensive roles could fundamentally change the nature of warfare.

From an ethical perspective, using AI in offensive operations raises concerns about accountability, proportionality, and the protection of civilians. If AI systems are granted increasing autonomy in target selection and engagement, assigning responsibility for their actions becomes challenging. Moreover, the ability of AI to process vast amounts of data and make rapid decisions could lead to unintended escalation or disproportionate responses, potentially violating the principles of just war theory.

Legal frameworks that govern the use of force, like international humanitarian law, will need to change to address the challenges posed by AI. The principles of distinction, proportionality, and military necessity, which form the cornerstone of IHL, may be challenging to apply to AI systems with varying degrees of autonomy. Additionally, the a◎ribution of a◎acks conducted by AI systems and the establishment of liability for any violations of international law will require careful consideration.

From a strategic standpoint, the offensive use of AI could alter the balance of power and the nature of deterrence. The scale at which AI systems operate may compress decision-making timelines, increasing the risk of miscalculation or inadvertent escalation. Moreover, the proliferation of AI-enabled offensive capabilities could lower the threshold for conflict, as states may perceive a strategic advantage in striking first.

We must ensure that humans control essential decisions, like choosing and engaging targets. Keeping AI systems transparent, easy to understand, and accountable is crucial. Working together globally to build trust and create rules for arms control could help manage the risks of using AI in offensive operations.

AI's ethical, legal, and strategic implications in offensive operations demand careful consideration and proactive governance. Balancing the potential military advantages with the imperative to uphold ethical principles, legal obligations, and strategic stability will be a defining challenge as AI continues to shape the future of warfare.

10.2 AI Tools in the Attacker's Arsenal

AI has become both a benefit and a challenge in the fast-changing world of cybersecurity. While AI-powered tools offer immense potential for enhancing security measures and defending against threats, they have become increasingly prevalent in the attacker's arsenal. Among the most concerning developments are deepfakes, AI-driven phishing campaigns, and the deployment of autonomous attack drones.

Deepfakes, AI Phishing, and Autonomous Attack Drones

Deepfakes: Deepfakes, which are highly realistic digital manipulations of audio and video content, have taken the world by storm. Attackers can use advanced machine learning algorithms to create realistic fake images and audio or video recordings of individuals, including public figures and high-profile targets. Deepfakes can spread disinformation, manipulate public opinion, or even blackmail individuals. The sophisticated nature of deepfakes makes them difficult to detect, posing significant challenges for cybersecurity professionals and society.

AI-Powered Phishing Attacks: Traditional phishing campaigns employ a broad, untargeted approach. Scammers send out mass emails with generic content to a large audience, hoping to deceive a portion of recipients. The goal is to trick individuals into inadvertently disclosing personally identifiable information, such as login credentials or financial details, or downloading malicious so◎ware (malware) onto their devices.

These phishing a◎empts rely on social engineering tactics, exploiting human trust and manipulating people into action. The more emails sent out, the higher the probability that some recipients will fall victim to the scam. While not highly sophisticated, traditional phishing remains prevalent due to its low cost and the potential for success through sheer volume.

Autonomous Attack Drones: The introduction of autonomous a◎ack drones has added a new layer to cybersecurity threats. Armed with AI, these drones can target critical infrastructure like power grids, communication networks, and industrial facilities. Their ability to operate independently, adjust to changing conditions, and make real-time decisions using AI algorithms makes them especially difficult to detect and stop. These drones' potential for physical damage and disruption underscores the need for counter-drone solid technologies and comprehensive defense strategies.

Combating AI-Driven Threats

Cybersecurity experts must adopt a proactive and multi-layered approach to combat these AI-driven threats. This approach includes investing in AI-powered defense mechanisms that respond to emerging threats in real-time. Machine learning algorithms can train to identify pa◎erns and anomalies associated with deepfakes, AI-generated phishing emails, and autonomous a◎ack drones. Additionally, organizations should also focus on providing the right education and awareness programs that will help individuals to identify and report suspicious activities.

Collaboration and Regulation

Collaboration between industry, government, and academia is crucial in developing effective countermeasures against AI-driven threats. The development of strong defense strategies can be accelerated if there is proper sharing of knowledge, best practices, research findings, and threat intelligence. Furthermore, establishing international frameworks and regulations governing the use of AI in cybersecurity can help mitigate the risks associated with malicious AI applications.

10.3 Developing Offensive AI Capabilities

The development of offensive AI capabilities has become a top priority for both governmental and private sector entities as they seek to gain a competitive edge in the increasingly evolving landscape of cybersecurity. By harnessing the power of AI, these organizations aim to create sophisticated tools and techniques that can automate and enhance various aspects of cybera◎acks.

Governmental and Private Sector Roles in Offensive AI

Governments worldwide are keenly interested in developing offensive AI capabilities, with military and intelligence agencies leading the way. They allocate significant resources to research and development efforts, o◎en collaborating with academic institutions and defense contractors to push the boundaries of what is possible with AI-powered cyber weapons.

Meanwhile, the private sector also plays a crucial role in developing offensive AI. Cybersecurity firms and technology companies invest heavily in AI research to create advanced tools and services for their clients, such as AI-powered vulnerability scanners, automated penetration testing platforms, and intelligent threat detection systems.

The private sector's involvement in offensive AI presents critical ethical issues. Companies must establish strict guidelines and oversight to prevent the exploitation of their AI-powered offensive tools.

Training AI for Offensive Cyber Operations

Organizations must invest in comprehensive training programs to effectively develop and deploy offensive AI capabilities. Training programs expose AI systems to vast amounts of real-world data, including network traffic, malware samples, and social media content, allowing them to learn and adapt to various scenarios.

Researchers use simulated environments and adversarial training techniques to prepare AI algorithms for offensive cyber operations. By subjecting AI to a wide range of simulated a◎acks and defenses, they can refine their models and improve their performance in real-world situations. This iterative training, testing, and refinement process is essential for developing robust and effective offensive AI capabilities.

10.4 Countering AI Offensive Measures

Developing effective countermeasures to mitigate their potential impact becomes paramount as offensive AI capabilities advance. One promising approach involves leveraging AI to counter offensive and misinformation campaigns.

AI as a Tool for Counter-Offensives and Misinformation Campaigns

Organizations can employ AI to detect and respond to AI-driven a◎acks in real-time. By si◎ing through extensive data and identifying pa◎erns and anomalies, AI-powered defense systems can quickly identify and neutralize offensive AI threats, such as AI-generated malware, fake social media accounts, and autonomous a◎ack vectors.

To effectively counter AI offensive measures, organizations must invest in the research and development of advanced AI algorithms specifically designed for defensive purposes. Advanced AI algorithms include machine learning models trained to recognize and classify AI-driven a◎acks and deep learning techniques for generating realistic decoy data to deceive and misdirect offensive AI systems.

AI also plays a vital role in countering misinformation campaigns, which have become increasingly prevalent in the digital age. By leveraging machine learning and natural language techniques, AI algorithms can identify and flag potential sources of disinformation, such as fake news articles, manipulated images, and deepfake videos.

Once identified, these AI-powered systems can help disseminate reliable and accurate information to counter the spread of misinformation through targeted content recommendations, automated fact-checking, and the generation of persuasive counter-narratives.

However, using AI for counter-offensives and misinformation campaigns also raises ethical concerns. Strong moral principles and oversight mechanisms must guide the development and deployment of these technologies to prevent their misuse and ensure transparency and accountability.

10.5 The Future of Offensive AI in Cyber Warfare

As AI continues to advance, it significantly impacts the future of cyber warfare. Integrating AI into offensive cyber operations can revolutionize how conflicts are waged in the digital domain, introducing new strategies, tactics, and capabilities that will shape the battlefields of tomorrow.

Predictive Analytics for Cyber Offense

One of the most promising applications of AI in offensive cyber operations is predictive analytics. By analyzing enormous data, AI systems can recognize patterns and trends in network traffic, user behavior, and system vulnerabilities. These AI capabilities enable the accurate prediction of potential targets and attack vectors.

This predictive capability allows offensive AI to stay one step ahead of defenders, adapting to changing circumstances and exploiting weaknesses in real-time. By continuously learning and evolving, these AI-powered offensive tools can become increasingly effective over time, posing a significant challenge to traditional cybersecurity measures.

Moreover, predictive analytics can anticipate and counter defensive measures, such as deploying new security patches or implementing more stringent access controls. By simulating various scenarios and testing different strategies, offensive AI can identify the most effective approaches to bypass or neutralize these defenses, increasing the chances of a successful attack.

The Impact of Quantum Computing on AI-Augmented Cyber Warfare

Another factor that will shape the future of offensive AI in cyber warfare is the advent of quantum computing. The subject of quantum computers is near and dear to the authors' hearts. While AI is a disruptive technology in and of itself, quantum computing is also a disruptive technology. We see two major technology disruptions happening almost simultaneously, so quantum computing holds a special place in the authors' hearts. As quantum computers become sufficiently powerful and accessible, they have the potential to revolutionize the field of AI, enabling the development of even more sophisticated and effective offensive tools.

Quantum computing can significantly speed up AI algorithms, le◎ing them complete complex calculations much faster than traditional computing methods. Increased computational power can enable offensive AI to analyze network traffic, identify vulnerabilities, and generate a◎ack scenarios at an unprecedented scale, overwhelming defensive measures and increasing the likelihood of successful breaches.

Furthermore, quantum computing can break traditional encryption methods, rendering many current cybersecurity measures obsolete. This quantum computing ability presents a significant challenge for defenders, who must develop new and innovative approaches to protect against quantum-powered offensive AI.

Combining AI and quantum computing creates a formidable force in cyber warfare. Quantum-enhanced AI algorithms can explore an exponentially more extensive set of possibilities and solutions, enabling them to identify and exploit vulnerabilities that may have been previously undetectable. This synergistic relationship between AI and quantum computing has the potential to reshape the cyber ba◎lefield, giving a◎ackers a significant advantage over defenders.

As quantum computers become more readily available, nation-states and non-state actors will likely invest heavily in developing quantum-powered offensive AI tools. The increased availability of quantum computers could lead to a new arms race in the digital domain as adversaries compete to gain a strategic edge through these advanced technologies.

Cybersecurity researchers and practitioners must begin exploring and developing quantum-resistant cryptographic methods and defensive strategies to counter this threat. Developing quantum-resistant cryptographic methods will require significant investment in research and development and close collaboration between academia, industry, and government agencies.

Moreover, developing quantum-powered defensive AI tools will be crucial in leveling the playing field against quantum-enhanced offensive AI. By harnessing the power of quantum computing for defensive purposes, such as identifying and mitigating quantum-based a◎acks, defenders can work to stay ahead of the evolving threat landscape.

Quantum computing significantly impacts AI-augmented cyber warfare. It represents a paradigm shi◎ in how we approach offensive and defensive operations in the digital realm. As we move forward, we must prioritize the development of quantum-resistant security measures and continue exploring the potential applications of quantum computing in AI and cybersecurity. We want to maintain a secure and resilient digital environment in the face of an ever-evolving threat landscape only by staying ahead of the curve and adapting to these emerging technologies.

10.6 Regulatory and Control Measures

As the use of AI in offensive cyber operations becomes more prevalent, the need for effective regulatory and control measures becomes increasingly important. Without proper oversight and governance, developing and deploying offensive AI tools could lead to unintended consequences, escalating conflicts and causing significant harm to individuals, organizations, and nations.

International Laws and Agreements on the Use of AI in Cyber Offense

The international community faces a significant hurdle in controlling the use of AI for cyberattacks. Current international laws and treaties often struggle to match the speed of rapid technological advancements, making it challenging to regulate AI-enhanced cyber warfare effectively.

To tackle this challenge head-on, countries worldwide must collaborate to create comprehensive international laws and agreements that directly address the distinctive issues posed by offensive AI. This effort should involve establishing clear rules for developing, testing, and using offensive AI tools and outlining the roles and responsibilities of both state and non-state actors participating in these operations.

Furthermore, global cooperation and information sharing will be crucial to ensure all involved parties can access the most current intelligence on emerging threats and the most effective strategies for reducing the dangers associated with offensive AI.

The Role of AI in Cybersecurity Policy Formation

In addition to international laws and agreements, AI can play a crucial role in shaping cybersecurity policies and regulations. By analyzing vast amounts of data on cyber incidents, threat patterns, and the effectiveness of various defensive measures, AI can provide valuable insights and recommendations to policymakers and regulators.

For example, AI algorithms can identify gaps and weaknesses in existing cybersecurity frameworks, highlighting areas that may require additional resources or regulatory interventions. Similarly, AI can simulate the potential impact of proposed policies and regulations, allowing decision-makers to assess their effectiveness and identify possible unintended consequences before implementation.

By leveraging the power of AI in policy formation, governments and organizations can develop more effective and adaptive regulatory measures that keep pace with the rapidly evolving landscape of offensive AI in cyber warfare.

10.7 Case Study: Offensive AI Deployed

Let us explore a real-world example to understand how offensive AI tools are used in cyber warfare.

Analysis of a Cyber Offensive that Utilized AI Tools

In 2021, a major cybersecurity firm reported on a sophisticated cyber offensive campaign that utilized AI-powered tools to target various government and private sector organizations across multiple countries. The a◎ackers employed a combination of AI-generated phishing emails, automated vulnerability scanning, and self-propagating malware to infiltrate networks and exfiltrate sensitive data.

One key aspect of this offensive involved using AI-powered social engineering tactics. The a◎ackers generated compelling phishing emails that mimicked trusted contacts' writing styles and communication pa◎erns by analyzing publicly available information on targeted individuals. This strategy allowed them to bypass traditional email security measures and trick recipients into revealing login credentials or installing malicious so◎ware.

Once inside the targeted networks, the a◎ackers deployed AI-powered malware to evade detection by learning and adapting to each system's security configurations. The malware was able to identify and exploit vulnerabilities in real-time, escalating privileges and moving laterally across the network to maximize the scope of the damage.

The a◎ack campaign stood out for its high level of automation and adaptability. AI-powered tools operate autonomously, making decisions tailored to each target's circumstances. This approach allowed the a◎ackers to scale their operations and achieve significant success, compromising multiple organizations before detection and neutralization.

The case study highlights the potential dangers of offensive AI in cyber warfare, showing how a◎ackers can use these tools to conduct highly effective and difficult-to-defend-against a◎acks. It underscores the importance of developing robust defensive measures and international regulatory frameworks to mitigate the risks posed by these emerging technologies.

Conclusion

As we review the topics discussed in chapters 9 and 10, it's clear that the cybersecurity landscape is constantly engaged in a race to stay ahead. The rapid expansion of AI technologies presents enormous opportunities and substantial challenges, necessitating continuous updates to our defense strategies and a consistent commitment to ethical practices.

AI could completely change how we secure our digital resources. It allows us to create advanced tools to find, stop, and react to cyber dangers quickly and precisely. AI-driven solutions, such as automated threat intelligence gathering and analysis, adaptive network security, and intelligent incident response, provide a hopeful way forward in the ba◎le against cybercrime.

However, as we have witnessed, the same AI technologies that help us defend can also be turned into weapons by bad actors. The emergence of offensive AI capabilities, like autonomous attack drones, AI-generated phishing campaigns, and self-spreading malware, presents a significant challenge to traditional cybersecurity measures. As attackers keep improving and changing their tactics, defenders must stay alert and proactive to stay one step ahead.

To deal with the complex world of AI, we must prioritize responsible development and use of AI technologies. We need a strategy that combines technical innovation, policy development, and international cooperation. By establishing clear guidelines and best practices for ethical AI use in cybersecurity, we can reduce the risks of these powerful tools and ensure they benefit society.

Moreover, the future of cybersecurity hinges on our commitment to continuously update our skills. As threats evolve, so must our expertise and tactics. Keeping skills current is essential in the fast-moving world of AI. Certification programs in AI, cybersecurity, and data science provide a cost-effective and efficient way to stay relevant. These programs quickly deliver the skills needed, offering a swifter alternative to pursuing a degree. Although degrees are valuable, certifications are efficient for meeting the demands of today's job market.

The path forward lies in recognizing the unbreakable connection between AI advancements and the future of cybersecurity. By embracing the responsible use of these technologies and joining forces to develop effective strategies for defense and deterrence, we can create a digital future that is more secure and resilient for all. The decision is ours - will we allow the potential risks of AI in cybersecurity to hinder our progress, or will we seize its remarkable potential to safeguard our digital realm like never before? The destiny of our online security hangs in the balance, and it falls upon us to choose which course we will chart.

CHAPTER 11

Cybersecurity in the Age of IoT and AIoT

Introduction

This chapter examines two significant innovations: The Internet of Things (IoT) and the artificial intelligence of things (AIoT). IoT and AIoT have revolutionized the way we interact with technology. AIoT takes this revolution a step further by fusing AI with IoT. This dynamic fusion births smart systems that possess the ability to think and act independently. These systems are good at looking at information, spo◎ing pa◎erns, and figuring things out mostly on their own (Minerva, Lightfoot, & Li, 2023).

IoT and AIoT have transformed various industries, from healthcare and manufacturing to smart homes and transportation. However, the rapid growth of connected devices and the integration of AI has also introduced new cybersecurity challenges. As the number of IoT devices continues to skyrocket, the a◎ack surface expands, making addressing the security risks of these technologies crucial.

IoT devices, often lacking robust security features, are a prime target for cyberattacks. Exploiting these vulnerabilities, hackers can infiltrate networks without authorization, stealing private information and potentially wreaking havoc on critical real-world systems. The ramifications of such breaches are grave, spanning from privacy infringements and financial losses to significant disruptions in operations and services.

To counter these threats, organizations must take a proactive stance toward cybersecurity. Safeguarding IoT devices and their data necessitates the implementation of robust security measures, including encryption, authentication, and access controls. Regular security audits and firmware updates are crucial to identifying and rectifying vulnerabilities before hackers can exploit them. By doing these things before problems happen, organizations can strengthen their security.

Integrating AI in IoT systems, known as AIoT, presents both opportunities and challenges for cybersecurity. On one hand, AI can enhance the security of IoT devices by enabling advanced threat detection, real-time monitoring, and automated incident response. However, hackers can also use AI to create smarter attacks that are harder to spot. Organizations must stay alert and constantly update their security to defend against these AI-powered threats.

We will talk about the main security problems that come with these technologies. We shall explore the best ways to reduce risks and keep everything safe. By learning about the threats and putting strong security measures in place, organizations can use IoT and AIoT to their advantage. At the same time, they will keep their systems and data safe and sound.

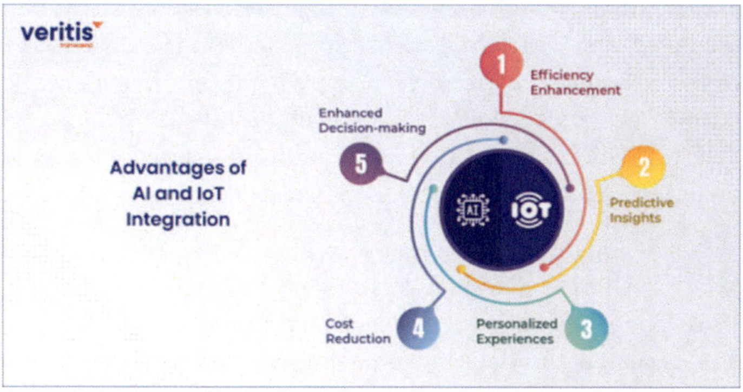

https://www.veritis.com/blog/ai-and-iot-collaboration-in-addressing-industry-challenges/

11.1 AI-Powered IoT Security Solutions

11.1.1 Threat Detection Using Machine Learning

Machine learning (ML) has become a vital weapon for making IoT security be◉er. Organizations can use ML algorithms to spot and stop cybera◉acks as they happen. These algorithms look at the enormous amounts of data that IoT devices create. They hunt for pa◉erns and things that seem strange, which could be signs of an a◉ack.

ML threat detection has a significant benefit - it can keep learning and changing over time. When new threats show up, and a◉ackers change their tricks, we can teach ML models using new data to catch threats we have not seen before. Being able to adapt is super important in the fast-moving world of cybersecurity. Old ways of detecting a◉acks by looking for specific signs o◉en miss new types of a◉acks.

You can deploy ML-based threat detection systems at various levels of the IoT ecosystem, from individual devices to gateway nodes and cloud platforms. By monitoring network traffic, device behavior, and user activities, these systems can detect a wide range of threats, including malware infections, unauthorized access attempts, and data exfiltration.

Organizations must pick and train their models carefully for ML-based threat detection to work well. They must gather a wide range of data that represents the problem well. They must also clean up the data by removing misleading or incorrect information. Finally, they should select the suitable algorithms and settings for the job. Regular testing and validation are also essential to assess the performance of ML models and fine-tune them as needed.

11.1.2 Real-Time Security Analytics for IoT Devices

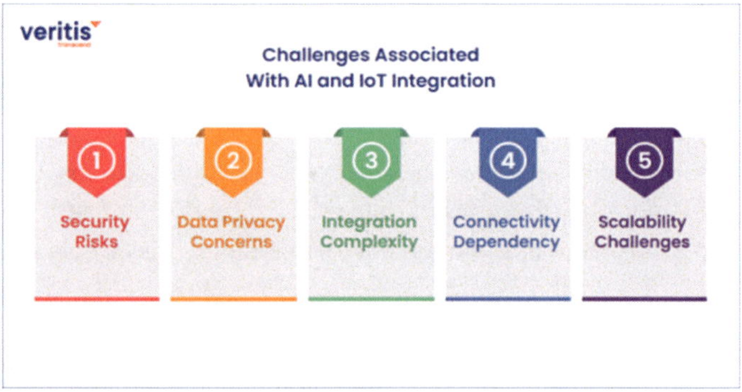

https://www.veritis.com/blog/ai-and-iot-collaboration-in-addressing-industry-challenges/

Real-time security analytics play a significant role in keeping IoT devices safe from cyberattacks. Organizations can watch and analyze the data that IoT devices create nonstop. This practice helps them see how secure they are and spot possible threats right away.

IoT devices generate a lot of data, such as sensor readings, network records, and user interactions. Real-time security analytics tools gather and process this data almost instantly. They use smart analysis methods to look for patterns and weird things that could indicate a security problem.

Real-time security analytics have a significant advantage - they can spot threats and let you tackle them fast. When you catch bad things happening as they occur, you can immediately jump into action to limit the danger and keep any damage to a minimum. This is super important with IoT. A hacker getting into an IoT device can cause problems like messing up critical systems or leaking private information.

Real-time security analytics solutions often leverage a combination of rule-based and machine learning approaches. Rule-based systems rely on predefined rules and thresholds to identify known threats, while machine learning algorithms can detect previously unknown threats by learning from historical data and adapting to new patterns.

Organizations must establish a robust data collection and processing infrastructure to implement effective real-time security analytics for IoT devices. This infrastructure involves deploying sensors and agents on IoT devices to collect relevant data, setting up secure communication channels to transmit the data, and establishing data storage and processing capabilities to handle IoT data's high volume and velocity.

Additionally, organizations must ensure that their real-time security analytics solutions are scalable and can handle the growing number of IoT devices in their ecosystems. Cloud-based platforms and distributed architectures can address these scalability challenges, enabling organizations to process and analyze IoT data at scale.

By combining threat detection using machine learning and real-time security analytics, organizations can establish a comprehensive and proactive approach to IoT security. These

AI-powered solutions enable organizations to identify and respond to threats quickly, minimizing the risk of successful cyberattacks and ensuring the integrity of their IoT ecosystems.

11.1.3 AI-Enhanced Encryption Techniques

AI has revolutionized the field of encryption, introducing innovative techniques that enhance the security and efficiency of data protection. AI-enhanced encryption leverages the power of machine learning algorithms to create more robust and adaptable encryption systems.

AI-enhanced encryption algorithms examine large quantities of data to recognize patterns and irregularities. The algorithms' ability to identify patterns and irregularities allows them to spot potential security risks and act immediately (Singh et al., 2023). AI-enhanced encryption remains at the forefront by constantly learning and developing. Continuous learning and development allow AI-enhanced encryption to adapt to new challenges and maintain the highest level of security.

One critical advantage of AI-enhanced encryption is its ability to generate dynamic and context-aware encryption keys. Traditional encryption methods rely on fixed keys that can become vulnerable over time. On the other hand, AI algorithms possess the capability to produce distinct keys for each data exchange. The algorithms consider several elements when generating the keys, including the nature of the data, the individuals or entities involved, and the present security environment. The dynamic key generation approach significantly increases the difficulty for attackers attempting to breach the encryption and obtain unlawful access to confidential data.

Moreover, AI-enhanced encryption methods can automate the management and distribution of keys. As modern encryption systems become increasingly complex, managing keys manually becomes overwhelming. AI algorithms can simplify critical management by automatically creating, sharing, and nullifying encryption keys when necessary. The automated key management process minimizes the chances of human mistakes and safeguards the reliability of the encryption system.

11.2 Securing AIoT Ecosystems

The fusion of artificial intelligence (AI) and the Internet of Things (IoT) has given rise to the AIoT ecosystem, opening unprecedented opportunities for innovation and automation (Lee & Kim, 2023). However, combining AI and IoT also brings about distinct security challenges that require attention to guarantee the integrity, confidentiality, and availability of AIoT systems.

11.2.1 Best Practices for AIoT Security

To effectively protect AIoT ecosystems, organizations need to employ a wide-ranging strategy that incorporates several best practices. One critical aspect is to implement robust authentication and access control measures. AIoT devices must use secure authentication techniques, like multi-factor authentication, to stop unauthorized individuals from gaining access. Access control policies should also be granular, granting users and devices only the necessary permissions to perform their intended functions.

Secondly, another crucial aspect is to encrypt data both when it is stored and when it is being transmitted. Encrypting data is necessary to safeguard sensitive information from being intercepted or tampered with. AIoT devices must utilize robust encryption algorithms and secure communication protocols. The use of strong encryption and secure protocols ensures the confidentiality and integrity of data as it moves between devices and backend systems.

Thirdly, regularly updating, and patching AIoT devices is critical to addressing known vulnerabilities and protecting against emerging threats. Manufacturers should provide timely firmware and software updates, and organizations should have processes in place; to ensure the prompt application of these updates across their AIoT ecosystem.

Fourthly, monitoring and logging the activities of AIoT devices can assist in the timely detection of and response to security incidents. Organizations must deploy centralized monitoring systems that gather and examine log data from AIoT devices. Centralized monitoring systems enable organizations to identify abnormal behavior and highlight any potential security breaches.

Frequent security assessments and penetration testing are crucial for pinpointing vulnerabilities and weaknesses in AIoT systems. The assessments must encompass the devices and the underlying infrastructure, such as networks, servers, and cloud platforms.

11.2.2 Challenges in Integrating AI with IoT Security

While AI can potentially enhance IoT security, integrating these technologies presents several challenges. One significant challenge is many IoT devices limited computational power and memory. AI algorithms often require substantial processing capabilities, which may not be available on resource-constrained IoT devices. This limitation can hinder the deployment of advanced AI-based security solutions directly on the devices themselves.

Another challenge is the heterogeneity of IoT devices and protocols. The IoT ecosystem comprises various devices from different manufacturers that run various operating systems and use diverse communication protocols. This heterogeneity makes it difficult to develop and deploy uniform AI-based security solutions that work seamlessly across all devices.

Data privacy issues present a challenge when combining AI with IoT security. AI algorithms often need large volumes of data to learn and accurately predict outcomes. However, collecting and processing data from IoT devices can raise privacy concerns, mainly when dealing with sensitive personal information. Organizations must balance leveraging data for security purposes and protecting user privacy.

Furthermore, ensuring the explainability and transparency of AI models employed in IoT security can be difficult. Numerous AI algorithms, especially deep learning ones, function as "black boxes." The "black box" nature of these algorithms makes it challenging to comprehend how they reach their conclusions. This lack of explainability can hinder the trust and adoption of AI-based security solutions in the IoT ecosystem.

11.2.3 Future Trends in AIoT Security Implementation

As the AIoT ecosystem continues to evolve, several trends are shaping the future of AIoT security implementation. One prominent trend is the increasing use of edge computing for AI-based security. Edge computing moves processing capabilities nearer to the IoT devices, enabling the implementation of AI algorithms directly on the devices or at the network edge. The edge computing approach decreases latency, enhances real-time threat detection, and reduces the need to transmit sensitive data to the cloud.

Another rising trend is the use of blockchain technology to bolster AIoT security. Blockchain offers a decentralized and tamper-resistant record-keeping system that organizations can utilize to store IoT data and confirm its accuracy. By integrating blockchain with AI, organizations can create secure and transparent AIoT ecosystems. The combination of blockchain and AI ensures the provenance and integrity of the data within these ecosystems.

Developing lightweight AI algorithms designed explicitly for resource-constrained IoT devices is also a growing trend. These algorithms optimize the trade-off between performance and resource usage, enabling the deployment of AI-based security solutions in a broader range of IoT devices.

Furthermore, the integration of AI with zero-trust security architectures is gaining traction. Zero-trust security assumes no device or user should be trusted and thus requires every user to provide continuous authentication and authorization. AI can enhance zero-trust security by enabling real-time risk assessment, behavioral analytics, and adaptive access control policies based on each device and user's context and risk profile.

Creating explainable AI (XAI) techniques is gaining significant traction in AIoT security (Wachter et al., 2020). XAI tries to make AI models easier to understand and explain. XAI lets users see the logic behind AI algorithms' decisions. By incorporating XAI techniques, organizations can foster trust and confidence in AI-based security solutions, ultimately promoting their wider adoption within the AIoT ecosystem.

Conclusion

The integration of AI with the Internet of Things (IoT), creating AIoT, is revolutionizing our interaction with technology. The combination of AI and IoT enables systems to be smart, self-sufficient, and able to make choices with li◎le human involvement. While AIoT offers transformative benefits across various sectors, such as healthcare and manufacturing, it also introduces complex cybersecurity challenges, especially as the number of connected devices continues to grow rapidly.

IoT devices, often equipped with inadequate security features, are vulnerable targets for cyberattacks. Successful attacks may result in unauthorized entry, information theft, and significant interruptions to services and infrastructure. To combat these threats, organizations must adopt proactive cybersecurity strategies. These include implementing robust security measures such as strong encryption, authentication, access controls, and regular security audits to identify and mitigate vulnerabilities.

ML has become a crucial tool for enhancing IoT security, enabling real-time detection and response to cyberthreats. ML algorithms can detect unusual patterns that may signify a security breach by examining the enormous quantities of data produced by IoT devices. ML's adaptive nature enables it to constantly update its knowledge base with new information, making it a powerful defense against evolving cyberthreats.

Real-time security analytics are crucial in monitoring IoT devices and initiating immediate responses to detected threats. This rapid action is essential to prevent minor incidents from escalating into more significant breaches. Organizations must have robust data management infrastructures to support real-time analytics, including adequate data collection, secure data transmission, and powerful processing capabilities.

AI has also significantly enhanced encryption techniques. AI-enhanced encryption systems can dynamically generate context-aware encryption keys, offering a more secure method of protecting data. These systems also automate key management, reducing human error and increasing encryption reliability.

However, securing AIoT ecosystems involves addressing unique challenges, such as integrating AI into diverse, often resource-constrained IoT devices. The integration must manage the extensive data requirements and privacy concerns associated with AI and IoT technologies. Ensuring the scalability and explainability of these systems is crucial for maintaining security and building user trust.

As AIoT grows, organizations must implement comprehensive and adaptive security strategies to harness its benefits while mitigating potential risks. These strategies must incorporate advanced technologies and best practices to safeguard against increasingly sophisticated cyberthreats.

Chapter 12

Cybersecurity Workforce in the AI Era

This generation believes that artificial intelligence technology is a popular fancy technical term, that has relevance in fictional movies and books. But in the ever-evolving world of cybersecurity, AI has been deeply integrated into several security measures. We currently live in an AI-powered world and therefore, it has become mandatory for cybersecurity professionals to be proficient in the latest AI tools and technologies so that they can effectively perform their regular duties. Now professionals cannot solely rely on firewalls and their traditional incident response expertise and must know data analysis, AI-powered threat intelligence, and even the ethical considerations related to the use of AI in their everyday work.

In the following sections, we will explore how universities, organizations, and individuals can adapt to the evolving cybersecurity world and remain updated to stay relevant in the industry and not be replaced by intelligent machines.

12.1 Skillsets and Training for AI-Integrated Cybersecurity Professionals

Firewalls, Intrusion Detection Systems, Incident Response systems, etc. are some of the tried and tested tools that will always be the priority in skills to learn. But as the world advances further, and the introduction of AI across all industries is becoming prevalent, it is time for cybersecurity professionals to invest in themselves and learn the new and advanced skills that will help them perform their jobs more effectively.

In the previous chapters, we learned how AI can be leveraged in various kinds of cybersecurity operations. So, learning about AI will also help cybersecurity professionals to use data analysis and figure out how the AI technology is working to predict and combat these cyberthreats.

They must know about the foundational AI knowledge, have basic data visualization skills, and ability to use security automation tools and platforms. This will help to enhance collaboration between humans and advanced intelligent machines in the future.

12.1.1 Foundational AI Knowledge

Foundational AI Knowledge refers to the fundamental understanding of AI technology that every cybersecurity professional must possess including the core concepts, algorithms, machine learning techniques, etc. Having this knowledge will help them utilize and interpret AI-powered cybersecurity tools effectively.

Here are some of the fundamental concepts they need to be aware of:

Core AI Concepts: Understanding the basic concepts of artificial intelligence like basics of machine learning, neural networks, deep learning, etc. Cybersecurity professionals must be good at understanding the underlying principles that drive AI technologies to assist in informed decision-making.

Algorithms: Should have a clear understanding of AI algorithms which involves knowledge of supervised and unsupervised learning, reinforcement learning, and other algorithmic approaches. Since every algorithm is useful in performing different sets of applications, cybersecurity experts must be clear about which one to select and apply for their tasks.

Machine Learning Techniques: Machine learning is a subset of artificial intelligence and it is necessary for cybersecurity professionals to be acquainted with different machine learning techniques including supervised learning like classification and regression, and unsupervised learning like clustering and reinforcement learning techniques.

This AI foundational knowledge becomes more significant as it enables professionals to efficiently use AI-powered cybersecurity tools for their everyday work. In the coming years when AI technology will get even more advanced, it will be more broadly used to mitigate even newer kinds of cyberthreats boosting demand for cybersecurity professionals who have a solid understanding of AI concepts and can make informed decisions about deploying and configuring AI tools. Not only that, this foundational knowledge will also help them analyze and interpret the results generated by AI algorithms.

12.1.2 Data Analysis and Visualization

Data Analysis and Visualization is an important skill not confined to just data science professionals. Even cybersecurity professionals must have these skills to effectively handle and derive insights from large data sets that are generated by different types of security tools and sensors. The primary goal of this kind of analysis is to identify patterns and anomalies within the data that can help detect potential cyberthreats.

Here are some of the important components every cybersecurity professional should know about:

- ***Large Datasets:*** This refers to large datasets that are in millions of terabytes. They are continuously generated by security tools, sensors, and other types of monitoring systems. The data can consist of logs, network traffic information, system events, etc. The cybersecurity workforce must know how to handle the enormous amount of data that might seem overwhelming without the right skills and tools.

- ***Identifying Patterns:*** Data analysis involves analyzing large datasets to extract insights. It can help identify regularities and patterns that might help distinguish between normal and abnormal activities/behaviors within a system or network. It becomes necessary to understand what is normal behavior and what could mean a deviation. By using data analysis, it will become easier for cybersecurity professionals to easily identify anomalies and figure out if any malicious activity is happening or about to happen like unusual network traffic or unusual user behavior.

- ***Anomaly Detection:*** The easiest way to detect whether the organization is under attack or soon going to be, is by detecting anomalies. So, cybersecurity professionals cannot compromise with this skill, and having a strong understanding of data analysis techniques can make this job easier. Statistical methods, machine learning algorithms, etc. can be used to automatically detect anomalies even in large data sets. A◎er interpreting the results of data analysis, the professionals can use their expertise and determine if there is any cybersecurity risk or not.

- ***Visualization:*** With the help of data visualization, the reports and insights can become easily understandable even if it is from large and complex data sets. So, learning this skill will help them represent their insights using graphical forms like charts, graphs, and dashboards. It also allows them to understand the pa◎erns and anomalies intricately which further helps them make quicker decisions making it important in scenarios where timely response to threats is important.

12.1.3 Threat Intelligence and Analysis

Source: https://www.agari.com/wp-content/uploads/2021/05/cyber-threat-intel-blog-

Being a cybersecurity professional, it is essential to understand various kinds of cyberthreats that not only pose risks in the current business landscape; but can be detrimental in the future as well. Knowing emerging threats is also necessary. This is where threat intelligence and analysis come into play.

Cybersecurity professionals must systematically examine insights from threat intelligence derived from artificial intelligence sources. The process might involve understanding the evolving cybera◎ack tactics, techniques, and procedures (TTPs) that are employed by cybera◎ackers who also use AI technology to carry out their crimes.

Scrutinizing the threat intelligence derived from AI sources will help cybersecurity experts gain insights into different innovative and sophisticated strategies utilized in AI-powered cybera◎acks. Thus, analyzing this threat intelligence in time can help to anticipate and mitigate potential threats before they happen. Organizations, in this way, can strengthen their current defense system by using threat intelligence and analysis.

12.1.4 Security Automation and Orchestration

Source: https://www.agari.com/wp-content/uploads/2021/05/cyber-threat-intel-blog-

It is the process of implementing and managing platforms that use artificial intelligence technology to perform and automate different kinds of cybersecurity tasks called Security Automation and Orchestration (SOAR). These AI-powered platforms are a great way to multiply an organization's security efforts. The goal of using SOAR is to automate routine and repetitive tasks across various security operations that will help cybersecurity experts focus on more strategic and innovative works.

The entire process of security automation involves using AI algorithms and various types of scripts that execute some predefined tasks requiring less to no manual intervention. Activities like routine threat detection, data analysis, incident response, etc. can be automated and it enhances the organization's efficiency and response time.

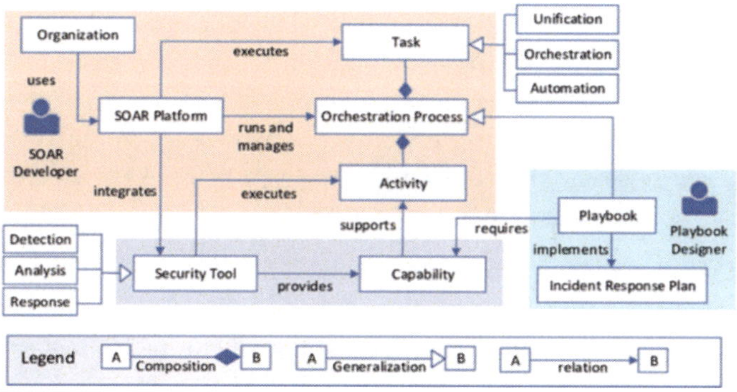

https://www.researchgate.net/figure/Conceptual-map-of-security-orchestration-and-automation_fig1_344260727

Therea⊚er, the coordination and integration of a variety of security tools and technologies available in a single framework is known as Orchestration. AI has an important role to play in this orchestration that leads to seamless communication and collaboration between various security systems, thus enabling a more comprehensive and holistic approach to cybersecurity.

If we are looking to prepare the workforce for the future, then it is highly important to equip them with these new skills. Universities and colleges can offer courses on AI for cybersecurity, data analysis, and ethical considerations and cybersecurity vendors can provide training programs specific to their AI-powered tools and platforms.

Those updating their cybersecurity skills with the latest technology and learning continuously will remain relevant to the industry and stay ahead of the curve in this rapidly evolving cybersecurity world.

12.2 Human-AI Collaboration in Cybersecurity

If we analyze properly, we can find AI is not a replacement for human professionals in the cybersecurity domain, instead, it is an incredible assistant. AI offers powerful tools that can help to make cybersecurity professionals' work easier as they can automate several tasks and perform them themselves without least human intervention.

However, it is the human skills required to handle powerful AI systems and tools. Therefore, the future of cybersecurity stands on an effective human-AI collaboration where the technology handles repetitive and heavy jobs like data analysis, threat detection, and response, and human analysts focus on more productive and innovative tasks like strategic decision-making, investigating incidents, and hunting threats.

For this model of human-AI collaboration to be effective, a new way of working is needed.

12.2.1 Defined Roles and Responsibilities

Defining a clear role and responsibility for both machines and humans is the best way to avoid any kind of confusion and ensure accountability within the cybersecurity workflow. Any organization seeking to integrate AI in their cybersecurity operations must establish the roles and responsibilities clearly which can include delegating specific tasks to both humans and AI systems, and some tasks specific to each of them.

This process also helps to ensure that each aspect of cybersecurity is handled by the right expertise and operates within its designated scope. While humans are be◎er for their contribution to contextual understanding, ethical considerations, and strategic decisions, AI systems can perform tasks like processing large datasets, identifying pa◎erns, and automating routine tasks at high speed and unmatched precision.

Defining roles and responsibilities also serves several purposes:

- It clearly outlines the work areas ensuring humans perform tasks requiring their expertise and AI does the jobs they can do best. This helps avoid confusion at work.
- It ensures accountability within the cybersecurity operations
- Defining clear roles and responsibilities can also assist in effective training and skill development measures. Effective training can be provided to humans to boost their creativity and make them more productive.

12.2.2 Trust and Transparency

When it comes to AI-human collaboration, trust and transparency cannot be compromised. Moreover, building trust in the AI tools the cybersecurity professionals will be using, and the insights and support the AI system will be providing, can help encourage collaboration. Also, these professionals should have a transparent view of the working of AI algorithms, and have a clear access and understanding of the datasets used to train the tools.

If professionals trust AI-powered cybersecurity tools, then they can easily rely on them for their cybersecurity work. They will have greater confidence in their work, accuracy, reliability, and ethical considerations. This makes it necessary for human analysts to have a clear understanding of how AI algorithms work, the methodologies they employ, and how the AI tools make decisions. In this context, transparency refers to providing humans with a clear knowledge of the workings of AI systems and ensuring that they are not perceived as "black boxes."

Moreover, transparency is not only limited to the understanding of the workings of AI algorithms but also extends to the data on which those algorithms are trained. The human analysts must know the nature and diversity of the datasets used for training along with the potential biases that might be present in the datasets.

If trust and transparency are maintained, it will help cybersecurity professionals to easily interpret and validate the outputs provided by AI tools. They can understand the limitations, and biases in a better way and know the areas where their judgment is necessary. Transparency can assist in bridging human trust with their AI counterparts and encourage building an environment where both humans and AI can contribute their strengths positively.

12.2.3 Effective Communication and Collaboration

Cybersecurity is already a complex field that needs real-time decision-making and prompt responses to different kinds of evolving cyberthreats. With effective human-AI collaboration, the overall efficiency of various cybersecurity measures can be improved.

Here are some reasons that indicate why effective communication and collaboration is necessary:

- It ensures the strengths and weaknesses of both human, as well as AI systems, are properly leveraged to combat different kinds of cyberthreats.
- It facilitates seamless interaction between tools and analysts that helps in the proper exchange of insights to assist in real-time decision-making
- • AI tools can generate insights rapidly and with high accuracy. However, it is the human expertise that can interpret those findings. So, with effective communication AI-generated insights are conveyed to the decision-makers.

12.2.4 Continuous Learning and Adaptation

There is no doubt that the cybersecurity landscape is continuously evolving and cyberattackers try to employ more innovative and sophisticated ways to exploit organization's resources. This makes it necessary for both human as well as AI systems to learn continuously and adapt to evolving forms of cyberthreats. The human-AI collaboration supports a continuous feedback loop where human analysts provide feedback to improve algorithms, and AI systems contribute by processing and learning from new data.

This process ensures both machines and humans continuously learn and stay adaptive and responsive.

12.2.5 Ethical Considerations

Ethical consideration is mandatory to effectively integrate AI in cybersecurity operations and it is human professionals that highlight the ethical considerations and judgment in the decision-making process. With effective communication and collaboration, the AI systems can be easily aligned with the ethical standards and the cybersecurity professionals can intervene whenever required to address any kind of ethical concerns arising. This collaborative approach promotes responsible and ethical use of AI in cybersecurity.

CHAPTER 13

AI Governance and Cybersecurity Policy

Throughout this journey, we understood the ever-evolving world of cybersecurity and the use of AI technology in this field. We saw how the integration of AI in cybersecurity can bring several possibilities as well as challenges. However, with the growing use of AI in this domain, several new challenges ought to come for the regulatory frameworks that require a comprehensive governance approach.

This chapter talks about the importance of AI governance and cybersecurity policy as well try to figure out complex consequences that can arise because of the relationship between AI and cybersecurity.

13.1 Regulatory Challenges in AI-Driven Cybersecurity

With the implementation of artificial intelligence (AI) to enhance cybersecurity operations, new and varied forms of challenges arise. As the AI field is highly dynamic, the traditional frameworks are no longer functional. This demands a recalibration of regulatory frameworks that match the pace of rapid technical advancements. With the increased importance of data and privacy, ethical considerations are highly talked about and features like transparency and accountability have become more important than ever. Therefore, the need for a striking balance between these technological advancements and ethical concerns has become necessary. In this section, we will scrutinize these challenges, and shed light on considerations related to data privacy, cross-border challenges, and adaptive regulations in the world of AI-powered cybersecurity.

13.1.1 A Different Set of Challenges in the Regulation of AI-Driven Cybersecurity

Lack of Clarity and Standardization

One of the most significant challenges is the lack of clarity and standardization in integrating AI into the cybersecurity domain. It is important to set a reliable and clear standard concerning different aspects of AI models including data quality, fair algorithms, their explainability, etc. However, the existing regulations often fall behind the rapid evolution of AI technology which always leads to a notable gap in regulations for developers and practitioners.

This lag in regulatory frameworks further increases the risk and uncertainty among the stakeholders on how to properly and ethically implement AI systems in cybersecurity. To address this challenge, it is necessary to adopt a regulatory framework that can match the pace of technological advancements along with providing a detailed and standard criterion for the evaluation of the trustworthiness of AI applications.

Algorithmic Bias and Discrimination

Many a time, the datasets on which the AI models are trained can consist of societal biases which can ultimately affect the decision-making process. Especially in cybersecurity applications like facial recognition, or risk assessment.

For example, facial recognition algorithms trained on biased data might result in unfair profiling, misidentification, or even exclusion of certain individuals or groups.

This necessitates the importance of addressing bias in training data and refining the AI models ensuring accurate and unbiased decisions.

Attribution and Liability

The complex question arises: who should be held responsible in the event of mistakes or misuse— the developer, the user, or the algorithm itself? This confusion demands a clear legal framework establishment that clearly defines the roles and responsibilities of these components. While developers can be held accountable for designing algorithms and ensuring security measures, the users should be accountable for employing AI tools responsibly, and algorithms for their actions.

Designing a clear and concise regulatory framework is important in properly assigning accountability, which can help address potential damages. There should be a perfect balance between encouraging innovations and establishing a foolproof framework that can define the accountability of relevant parties.

Cross-Border Regulatory Disparities

Cross-border regulatory disparities in the context of AI-driven cybersecurity refer to inconsistencies and differences in regulatory frameworks and standards across different nations. In a world where cyberthreats have become global, there is an absolute need for a collaborative and unified regulatory approach for increased cybersecurity efforts worldwide. It is another big challenge that might hinder the proper exchange of information, technologies, and collaborative efforts between different countries.

13.2 Global Cooperation for Cybersecurity and AI Governance

In this section, we will explore the growing importance of collaborative efforts at the international level to address the challenges arising from the integration of artificial intelligence into cybersecurity. It will focus on the united efforts including sharing threat intelligence to design a standard governance policy. We shall try to understand the need for diplomacy, standardization, and capacity building justifying the collaborative effort in mitigating cybersecurity challenges and AI governance.

Here are some ways in which Cybersecurity and AI Governance can be improved globally:

13.2.1 Shared Threat Intelligence

Shared threat intelligence refers to the practice of exchanging information and insights regarding cybersecurity threats among nations, organizations, and cybersecurity experts on a global level. Cybersecurity is a global issue and is a borderless threat, therefore, it requires a collaborative and collective effort from all the countries in the world. Sharing threat intelligence is an important strategy that will help governments and organizations understand emerging threats, vulnerabilities, and attack patterns on a global level. Thus, there is a need to establish a set mechanism for threat intelligence sharing. With smooth and efficient sharing of information, a coordinated response can be established among countries for potential cyberthreats which will help in the early detection and elimination of cyber risks.

This collaboration and sharing of information will encourage to building of a more secure and safer cybersecurity landscape that can leverage the expertise and insights from across the globe.

13.2.2 Developing International AI Governance Frameworks

This refers to the creation and adoption of a set of principles and guidelines that will act as a standard for the development and deployment of AI globally.

Some popular frameworks that exist include:

- OECD AI Principles
- EU Ethics Guidelines for Trustworthy AI
- UNESCO Recommendation on the Ethics of Artificial Intelligence.

These are great initiatives providing a comprehensive foundation for ethical and responsible AI practices giving more emphasis on values such as transparency, accountability, fairness, and human-centric design.

Now, when it comes to cybersecurity, adhering to these international frameworks will ensure that AI tools designed and used for cybersecurity operations will prioritize security, privacy, and ethical considerations. Policymakers, organizations, and developers can use these frameworks as a guide which will help them follow a common understanding of ethical standards. And it will promote a cohesive approach to AI governance on a global level.

Establishing these international principles ensures a shared foundation that can be adopted and adapted by nations and organizations across the world.

13.2.3 Building Trust and Public Confidence

Building trust and public confidence in the application of artificial intelligence (AI) for cybersecurity requires collective efforts in public education and awareness campaigns. The AI technology is very complex and it often leads to misconceptions and assumptions among the general public. By having proactive communication between the concerned parties and the public, along with educational initiatives, the concerns related to the use of AI can be eliminated.

Public Education Campaigns are particularly effective in explaining the benefits, challenges, limitations, and safety measures related to the use of AI in cybersecurity. When relevant information is made accessible to the stakeholders, a be◎er understanding can be developed of how AI contributes to enhancing digital security and preventing cyberthreats.

Engaging the stakeholders and general public in an open dialogue can also encourage transparency and help them understand the role of AI in cybersecurity clearly. It will also encourage an exchange of ideas, information, concerns, as well as expectations. It will be particularly helpful in shaping the development of AI that can align with societal values and ethical standards.

By ensuring responsible AI development including ethical considerations, fairness, and accountability, public acceptance can be gained easily. The fact is simple, when people perceive that AI technologies are developed and deployed for their best interest only, trust will automatically build and they will be more confident in using AI for cybersecurity. So, it is essential to actively involve the public and maintain transparency. This will ensure there is a healthy and reliable relationship between society and AI technologies.

In conclusion, the use of AI in cybersecurity requires active governance where regulators must consider the evolving threats, rapidly advancing technologies, and public concerns while they address the ethical aspects of AI in cybersecurity. Encouraging international collaboration and standardizing the governance principles will help the world to collectively build a strong defense against emerging cyberthreats and pave the path toward a more secure digital future.

13.3 International Collaboration and Standards for Secure and Ethical AI

All through the previous chapters, we understood how deeply AI has become integrated into cybersecurity operations. This increased integration has made international collaboration and standardized practices highly important. Today, building and deploying secure and ethical AI is not just a national but global concern, as cyberthreats normally transcend borders.

In this chapter let us explore the importance of international cooperation and the development of global standards to ensure AI is used securely and ethically in the global cybersecurity environment.

13.3.1 The Need for International Collaboration

- **Global Nature of CyberThreats**

Cyberthreats are inherently global. That means cybercriminals often operate across different countries and target victims from across the world. Because of this, no single nation can effectively combat these threats alone. Therefore, international collaboration becomes mandatory to help with intelligence, resources, and best practices sharing. It makes a unified force to combat international cybercrimes.

- **Harmonization of Regulations and Standards**

Different countries have different regulations and standards for AI and cybersecurity. There is an absolute need for harmonization of these regulations to reduce conflicts and create a cohesive framework. It will ultimately help with secure and ethical AI deployment. This alignment helps to create interoperable systems and protocols that make it easier for countries to collaborate on joint cybersecurity initiatives.

13.3.2 Areas for International Collaboration

- **Information Sharing**

Countries must be ready to share information about cyberthreats, vulnerabilities, and emerging a◎ack pa◎erns. This knowledge sharing will help countries make effective defense strategies. International organizations and coalitions such as the Global Forum on Cyber Expertise (GFCE), and the Cybersecurity and Infrastructure Security Agency (CISA) play important roles in sharing of such information.

- **Joint Research and Development**

Collaborative research and development (R&D) efforts can maximize innovation in the field of AI and cybersecurity. By sharing expertise and resources, countries can accelerate the development of advanced AI tools and techniques required to detect and mitigate cyberthreats. With joint R&D, countries can also focus on addressing common challenges that come in the way of using AI for cybersecurity like Biases, and unexplainable AI models.

- **Capacity Building**

International collaboration also helps with building cybersecurity capacity even in less developed regions. Countries can collaborate to train professionals, offer resources, and provide technical assistance to less fortunate regions and help them strengthen their cybersecurity efforts. Capacity-building initiatives ensure all countries, irrespective of how developed they are, benefit from and contribute to secure and ethical AI practices.

13.3.3 Establishing International Standards

Importance of Standards

Standards serve as a framework that ensures the security, reliability, and ethical use of AI in cybersecurity. They provide proper guidelines and best practices for organizations to follow to develop and deploy AI systems responsibly. These international standards also facilitate interoperability and help different AI systems to work seamlessly across borders.

Existing Standards and Frameworks

Many organizations are working to develop standards for AI and cybersecurity. For example, the International Organization for Standards (ISO) and the International Electrotechnical Commission (IEC) have published standards like ISO/IEC 27001 for information security management and ISO/IEC JTC 1/SC 42 for AI. Such standards cover different aspects of AI development such as data quality, algorithms, risk management, etc.

Developing New Standards

As technology evolves, there is a continuous need to develop new standards that can address new and emerging challenges. Stakeholders including government, organizations, industry leaders, civil servants, and even normal citizens need to collaborate to identify gaps in current standards and work together to create comprehensive guidelines. AI ethics, bias mitigation, data privacy, security of AI systems, etc. These are some of the areas of focus to consider in new standards.

Conclusion

International collaboration and the establishment of global standards have become mandatory to ensure AI is used securely and ethically in the field of cybersecurity. Countries must work together and address some of the common challenges, share their valuable knowledge, and the best practices to create a cohesive framework that can be used to combat all kinds of global cyberthreats effectively. A collaborative effort will ensure global cyberspace is safe and protects privacy for everyone.

CHAPTER 14

Conclusion: Embracing the AI-Driven Cybersecurity Landscape

Throughout this book, we have embarked on a captivating journey exploring AI's inevitable and transformative integration into cybersecurity. The convergence of these two powerful technologies has reshaped how we approach digital security, offering unprecedented opportunities and complex challenges. As we traversed the chapters, we saw that AI is more than just a theoretical idea in academic journals or science fiction. Instead, it is a practical and essential tool in the fight against evolving cyberthreats.

From the fundamentals of AI and its application in cybersecurity to the advanced strategies and future implications, we have witnessed a remarkable transition from conceptual discussions to real-world applications. As seasoned technology leaders with extensive experience in both AI and cybersecurity, we have generously shared our invaluable insights and experiences, guiding readers through the intricacies of this dynamic and rapidly evolving landscape.

As we conclude this enlightening journey, let's reflect on the key takeaways and lessons learned. The knowledge and strategies shared in this book will provide a solid foundation for professionals, organizations, and individuals aiming to leverage AI in their cybersecurity efforts.

However, our journey does not end here. We must remain vigilant, adaptable, and proactive when confronting emerging challenges and opportunities as we enter the uncharted future. With the insights gained from this book as a guide, we can confidently navigate the complex terrain of AI-driven cybersecurity and prepare ourselves for the exciting and transformative times that lie ahead, ensuring that we are well-prepared for the future of cybersecurity.

14.1 Key Takeaways from the Book

This book explores three key themes in cybersecurity:

1. **Enhanced Threat Detection and Response:** Deep learning is revolutionizing the identification and combating of cyberthreats.
2. **Challenges of AI in Cybersecurity:** While AI offers immense potential, its implementation has hurdles to overcome.
3. **Ethical and Regulatory Considerations:** AI in cybersecurity raises important ethical and regulatory questions we must address.

The following sections delve deeper into each of these themes.

14.1.1 Enhanced Threat Detection and Response

The advancements in AI-driven threat detection have revolutionized how organizations defend against cyberattacks. Machine learning algorithms have become crucial tools, allowing security systems to process large volumes of data, recognize patterns, and detect unusual activities in real-time. These AI-powered solutions, such as next-generation firewalls and intrusion detection systems, have significantly enhanced the accuracy and speed of threat detection, allowing organizations to respond swiftly and mitigate potential damages. Throughout the book, we have highlighted numerous success stories where AI has been instrumental in thwarting sophisticated cyberthreats that would have otherwise gone unnoticed.

Moreover, the role of machine learning in anomaly detection has been particularly transformative. By leveraging behavioral analysis, AI algorithms can identify suspicious activities and potential insider threats, providing a proactive approach to cybersecurity. We have shared compelling case studies demonstrating how AI has helped organizations uncover hidden vulnerabilities and prevent data breaches before they occur.

We explore using deep learning techniques in cybersecurity, mainly for malware detection. Deep learning models have successfully identified malicious code patterns and variations, even when attackers use new methods to conceal them. These techniques highlight the power of AI in staying ahead of emerging threats.

By leveraging deep learning, cybersecurity systems can continuously adapt and improve their detection capabilities, offering robust protection against sophisticated cyberattacks. This approach enhances security and reduces the response time to new threats, ensuring a more resilient digital environment. As attackers evolve their strategies, the dynamic nature of deep learning models ensures that defense mechanisms remain one step ahead, providing a critical advantage in the ongoing battle against cyberthreats.

14.1.2 Challenges of AI in Cybersecurity

While the benefits of AI in cybersecurity are undeniable, we also illuminated the complexities and challenges associated with its integration. Addressing issues like algorithmic bias, privacy concerns, and the need for transparency is crucial to ensuring AI's responsible and ethical deployment in cybersecurity practices.

AI algorithmic bias can lead to false positives or negatives, compromising the effectiveness of security systems. We recommend mitigating bias through diverse training data and regular audits. Cybersecurity professionals must be mindful of these potential pitfalls and actively work to address them. Over-reliance on biased AI systems can lead to false alarms, unnecessary disruptions, and eroded trust in the technology. Regularly assessing and fine-tuning AI models ensures their accuracy and reliability. This practice is crucial for maintaining the effectiveness of AI-driven security systems and avoiding the negative consequences of biased algorithms.

In addition, we explored privacy concerns surrounding AI in cybersecurity. Robust data protection measures and clear data usage and storage guidelines are crucial. Since AI systems often need access to sensitive data to learn and improve, maintaining privacy is essential. This guidance highlights the delicate balance between leveraging AI for enhanced security and safeguarding individual privacy rights, providing recommendations for organizations to navigate this complex landscape.

Effective data governance involves setting clear policies, ensuring regulation compliance, and continuously monitoring AI systems for vulnerabilities. Organizations can protect sensitive information by prioritizing data governance and maintaining trust in their AI-powered security solutions.

14.1.3 Ethical and Regulatory Considerations

The ethical implications of AI in cybersecurity are a recurring theme throughout the book. The authors emphasize the importance of maintaining fairness, accountability, and transparency in deploying AI systems and advocate for developing ethical frameworks and guidelines to ensure that AI is used responsibly and aligned with societal values.

We make a compelling case for addressing the ethical challenges posed by AI in cybersecurity, such as biases or malicious uses. Cybersecurity professionals must ensure AI is developed and deployed ethically. Clear oversight and accountability mechanisms are essential.

The book explores the potential consequences of failing to address these ethical concerns. Biased AI systems in cybersecurity can lead to unintended discrimination, where specific groups or individuals may be disproportionately targeted or subjected to unfair treatment. For example, an AI-powered intrusion detection system trained on biased data may generate more false positives for specific demographics, leading to unfair scrutiny or wrongful accusations.

Moreover, the authors emphasize the need to establish clear accountability mechanisms for AI-driven decisions in cybersecurity. When an AI system makes a critical decision, such as blocking a suspected cybera◎ack or flagging a potential insider threat, there must be a way to retrace the decision-making process. This transparency builds trust in AI systems and ensures they operate fairly and unbiasedly.

The book stresses the importance of human oversight in AI-driven cybersecurity. While AI can automate tasks and provide valuable insights, it should only operate fully autonomously—at least not yet. Human experts must be involved in the loop, reviewing, and validating the decisions made by AI systems. This human-in-the-loop approach helps mitigate the risks of AI making erroneous or biased decisions and ensures that there is a level of accountability and responsibility in the deployment of AI in cybersecurity.

To address these ethical challenges, the authors recommend the development of comprehensive ethical frameworks specifically tailored to AI in cybersecurity. These frameworks should provide guidelines and best practices for designing, developing, and deploying AI systems in a manner that upholds ethical principles. They should cover aspects such as data privacy, bias mitigation, transparency, and accountability and provide a roadmap for organizations to navigate the ethical complexities of AI in cybersecurity.

The book also addresses regulatory challenges in AI cybersecurity, emphasizing the need for global cooperation and international standards. Effective AI governance requires collaboration between governments, academic institutions, and industry leaders. This cooperation is crucial for creating a regulatory environment that promotes innovation while ensuring responsible AI deployment.

In general, governments help shape the regulatory landscape for AI in cybersecurity. They must establish clear laws and regulations that provide a framework for developing and using AI in this domain. These regulations should balance fostering innovation and protecting individual rights and societal values. Governments should also invest in research and development initiatives to advance the responsible use of AI in cybersecurity and support the development of industry standards and best practices.

Academic institutions are essential in addressing AI's regulatory challenges in cybersecurity. They can research AI's ethical and legal implications, develop frameworks for responsible AI use, and help prepare the workforce for AI's opportunities and challenges. Integrating academia and industry can foster knowledge exchange and ensure that research findings lead to practical applications.

Industry leaders, including cybersecurity companies and technology providers, must also actively participate in shaping the regulatory landscape for AI in cybersecurity. They should dialogue with governments and academic institutions to share their insights and contribute to developing industry standards and best practices. By working together, industry leaders can help to create a level playing field for the responsible development and deployment of AI in cybersecurity.

Effective AI governance in cybersecurity requires ongoing collaboration and dialogue between all stakeholders. As AI systems evolve, new challenges emerge. We need a flexible and adaptive regulatory framework to keep up with these changes. Regularly reviewing and updating regulations, standards, and best practices will ensure they stay relevant and effective against evolving AI capabilities and cyberthreats.

14.2 The Future of AI and Cybersecurity

As we look toward the future, AI has immense potential to revolutionize cybersecurity. This book explores various technological advancements on the horizon, such as quantum computing and advanced predictive analytics, which can significantly enhance AI's capabilities in detecting and mitigating cyberthreats.

Quantum computing, for example, could enable AI algorithms to process vast amounts of data at unprecedented speeds, allowing for real-time threat detection and response. The authors briefly discussed the principles behind quantum computing and its potential applications in cybersecurity. They explain how quantum algorithms can solve complex optimization problems and perform rapid searches through massive datasets, which could significantly enhance the efficiency and accuracy of AI-driven security systems.

Advanced predictive analytics is another area where AI holds immense promise for cybersecurity. By leveraging techniques like deep learning and reinforcement learning, AI systems can analyze historical data, identify pa◎erns, and predict potential cyberthreats before they happen.

However, we also acknowledge that these advanced techniques bring new challenges and require continuous learning and adaptation. Cybersecurity professionals must keep up with the latest technologies and techniques as threats evolve and a◎ackers grow more sophisticated. We emphasize the strategic importance of AI cybersecurity training and developing a skilled workforce to use these cu◎ing-edge tools effectively.

The book provides valuable insights into the skills and training required to navigate the AI-driven cybersecurity landscape. It highlights the need for a multidisciplinary approach, combining cybersecurity, data science, and AI expertise. Cybersecurity professionals must understand AI algorithms, machine learning techniques, and data analysis methodologies to design, implement, and maintain AI-powered security systems effectively.

We stress the importance of so◎ skills like critical thinking, problem-solving, and effective communication. These essential skills help translate technical findings into actionable insights for decision-makers.

Despite their extensive knowledge and experience, cybersecurity professionals sometimes need help finding practical solutions. They can develop robust defense mechanisms by collaborating with AI experts and being willing to learn and adapt. This approach shows the value of embracing new technologies and the power of interdisciplinary collaboration in facing evolving cyberthreats.

While still in the early stages of development, autonomous cybersecurity systems have the potential to significantly reduce the workload on human cybersecurity professionals and provide round-the-clock protection against evolving threats. However, we caution against overreliance on AI in cybersecurity. While AI can automate tasks and enhance decision-making, more complete solutions exist. Effective cybersecurity strategies combine AI-driven automation with human expertise, requiring oversight and intervention. Cybersecurity professionals must remain vigilant, interpret AI systems' insights, and exercise critical thinking to make informed decisions.

Additionally, this book explores the potential impact of AI on the cybersecurity job market. Some experts worry that AI might replace human cybersecurity professionals, leading to job losses in the industry. However, the authors present a different perspective, contending that AI will primarily augment and enhance human capabilities rather than replace them entirely. We predict a shi◎ in the roles and responsibilities of cybersecurity professionals, with a greater emphasis on strategic planning, risk management, and the development of AI-powered solutions.

New job roles, like AI security specialists and generative AI prompt engineers, will design, implement, and maintain AI-driven cybersecurity systems. These professionals need a unique mix of technical skills, business acumen, and domain knowledge to bridge the gap between AI and cybersecurity effectively. The authors also stress the importance of upskilling and reskilling the current cybersecurity workforce to prepare them for the AI-driven future.

However, the book acknowledges AI's ethical and societal implications in cybersecurity. The authors raise important questions about privacy, data protection, and the potential for AI to be misused or exploited by malicious actors. We emphasized the need for responsible AI development and deployment, focusing on transparency, accountability, and fairness. The book calls for establishing ethical frameworks and guidelines to ensure that AI-driven cybersecurity solutions are developed and used in a manner that aligns with societal values and protects individual rights.

AI in cybersecurity offers immense opportunities and complex challenges. We explored the technological advancements, skills requirements, and ethical considerations shaping this evolving field. As we move forward, it is crucial to approach AI in cybersecurity with a balanced perspective, recognizing its potential while staying vigilant and adaptable to new challenges.

14.3 Preparing for the Uncharted Future

Organizations and professionals must adopt effective adaptation strategies as we navigate the uncharted future of AI in cybersecurity. We offered valuable recommendations for staying proactive and ensuring the security of digital assets and infrastructures.

We have also stressed the importance of collaboration and global initiatives in tackling AI cybersecurity challenges. Cyberthreats know no boundaries, and effective defense requires a coordinated effort across nations and organizations. The book has called for developing international standards and policies to ensure a unified approach to AI cybersecurity governance.

Preparing for the future requires proactive cybersecurity education and awareness. Organizations must invest in comprehensive training programs covering technical skills and ethical considerations related to AI. Raising public awareness about AI's benefits and risks is also crucial. This awareness motivates individuals to make informed decisions about their digital security.

The authors have explored the concept of responsible AI development in the context of cybersecurity. They have advocated for integrating ethical considerations into the foundation of AI systems, ensuring that these technologies are built with security, privacy, and fairness in mind from the ground up.

We also discussed the role of governments and policymakers in shaping AI's future in cybersecurity. Comprehensive national strategies must prioritize research, innovation, and collaboration. Governments should invest in public-private partnerships and international cooperation to tackle global cyberthreats.

14.4 Conclusion and Final Thoughts

As we conclude our exploration of AI in cybersecurity, it's clear that AI offers immense potential and significant challenges. Integrating AI can revolutionize threat detection, response, and prevention and introduce new complexities and risks.

The key takeaway is the importance of preparedness and proactive engagement with AI technologies. Organizations and cybersecurity professionals must embrace AI-driven solutions while considering ethical and regulatory aspects. Developing a deep understanding of AI's capabilities and limitations is crucial. Continuous adaptation and innovation are essential to face evolving threats.

Throughout the book, the authors have provided a comprehensive framework for navigating the AI-driven cybersecurity landscape. From the fundamentals of AI and cybersecurity to advanced strategies and future implications, readers have gained a solid foundation for understanding and leveraging AI in their cybersecurity practices. These pages' practical insights, real-world examples, and expert guidance will be valuable resources for professionals and organizations.

As we move forward, we must recognize that AI's journey in cybersecurity is far from over. The technology will continue to evolve, presenting new opportunities and challenges. It is up to us, as cybersecurity professionals, researchers, and decision-makers, to shape the future of AI in a manner that prioritizes security, privacy, and ethics. AI's success in cybersecurity ultimately depends on balancing its transformative potential with risk mitigation. This balance requires ongoing dialogue, research, and collaboration among all stakeholders to ensure that AI is developed and deployed responsibly and beneficially.

Let us embrace the opportunities ahead while staying alert to new challenges. We can build a future where AI helps create a secure and resilient digital world by working together. The journey may be complex and uncertain, but with the proper knowledge, skills, and mindset, we can navigate it confidently and emerge stronger.

14.5 Closing Remarks

We recognize the crucial role of cybersecurity professionals in leading responsible and effective AI integration. The future of cybersecurity depends on those who adapt to new technologies, champion ethical practices, and embrace change. These individuals will lead the way in shaping the AI-driven cybersecurity landscape. They will harness AI's benefits while mitigating its risks and challenges.

The evolution of cybersecurity is ongoing, and this book serves as a starting point for understanding AI's transformative potential. It offers a solid foundation and strategic direction for navigating future complexities and opportunities. However, the learning journey continues because as AI advances and new challenges arise, staying informed and proactive is crucial. Cybersecurity professionals must commit to continuous learning to stay ahead.

We thank the readers for embarking on this journey with us. We hope the insights and knowledge shared in these pages empower everyone to shape a secure and resilient digital future. This book aims to spark curiosity, inspire innovation, and foster a sense of responsibility among all its readers. It encourages everyone to think critically about AI's implications and actively participate in the discourse surrounding its development and deployment.

As we step into the uncharted territory of AI-driven cybersecurity, let us remember the words of the renowned cybersecurity expert Bruce Schneier: "Security is not a product, but a process." This powerful statement emphasizes that cybersecurity is not a one-time solution but an ongoing endeavor that requires continuous effort and adaptation. In the context of AI, we must approach its integration into cybersecurity as a dynamic and iterative process, constantly evaluating its effectiveness, addressing emerging challenges, and refining our strategies.

To successfully navigate the AI-driven cybersecurity landscape, professionals must embrace continuous learning, seek opportunities to enhance their skills, and actively collaborate with industry peers and experts. A◎ending conferences, workshops, and training programs, pursuing certifications in artificial intelligence, data science, and cybersecurity, and staying current with AI and cybersecurity advancements are crucial.

Additionally, professionals must be willing to think outside the box, explore new approaches, and creatively leverage AI technologies to combat evolving cyberthreats. This perspective may involve developing novel AI algorithms, designing intelligent security architectures, or finding innovative ways to integrate AI with existing cybersecurity tools and processes.

Collaboration is equally essential in the age of AI-driven cybersecurity. The challenges posed by AI are complex and multifaceted, requiring expertise from various domains, including cybersecurity, data science, ethics, and policy. We must actively collaborate across professional disciplines to develop holistic and practical AI solutions. This collaborative approach ensures AI development and deployment align with ethical principles, legal frameworks, and societal values.

As we navigate the exciting and challenging times ahead, let this book be a catalyst for positively embracing the integration of AI and cybersecurity. It calls readers to action, urging them to shape the future of AI in cybersecurity. We can harness AI's transformative potential by embracing continuous learning, innovation, and collaboration while ensuring our digital world's security and resilience.

The future may be uncertain, but with the knowledge and insights from this book, professionals are ready to handle challenges and take advantage of AI's opportunities. As we move forward, let us approach AI-driven cybersecurity with a sense of responsibility, curiosity, and determination. Together, we can build a future where AI and cybersecurity unite to protect our digital assets, defend against cyberthreats, and create a more secure and trustworthy digital ecosystem for everyone.

References and Further Reading

- https://medium.com/@neerseth8/are-you-afraid-of-ai-but-why-a386330f70ea
- https://technowastrip.com/blog/technology-3/navigating-the-complex-terrain-of-cyber-security-safeguarding-the-digital-realm-2
- https://www.hdfcergo.com/blogs/cyber-insurance/cyber-security-risks-for-professionals-in-2023
- https://gitnux.org/cybersecurity-in-the-ai-in-insurance-industry
- h◎ps://www.techradar.com/features/ai-has-opened-a-new-front-in-the-war-with-cybercrime
- https://www.bulbapp.io/p/98edfd7a-0309-4b2b-a65d-00119eb833f3/the-ethical-issues-of-artificial-intelligence
- h◎ps://www.secopsolution.com/blog/artificial-intelligence-on-cybersecurity
- https://www.balbix.com/insights/artificial-intelligence-in-cybersecurity/
- https://www.ncsc.gov.uk/guidance/ai-and-cyber-security-what-you-need-to-know
- https://www.securitymagazine.com/articles/100312-the-intersection-of-cybersecurity-and-artificial-intelligence
- https://www.checkpoint.com/cyber-hub/cyber-security/what-is-ai-cyber-security/

- https://www.crowdstrike.com/cybersecurity-101/machine-learning-cybersecurity/
- https://www.linkedin.com/posts/jeff-hughes-85b12810_ai-in-cyber-security-friend-or-foe-activity-7164257971500253184-Xh2o
- https://medium.com/@neerseth8/are-you-afraid-of-ai-but-why-a386330f70ea
- https://technowastrip.com/blog/technology-3/navigating-the-complex-terrain-of-cyber-security-safeguarding-the-digital-realm-2
- https://www.hdfcergo.com/blogs/cyber-insurance/cyber-security-risks-for-professionals-in-2023
- https://gitnux.org/cybersecurity-in-the-ai-in-insurance-industry
- https://www.techradar.com/features/ai-has-opened-a-new-front-in-the-war-with-cybercrime
- https://www.bulbapp.io/p/98edfd7a-0309-4b2b-a65d-00119eb833f3/the-ethical-issues-of-artificial-intelligence
- https://www.secopsolution.com/blog/artificial-intelligence-on-cybersecurity
- https://www.linkedin.com/posts/jeff-hughes-85b12810_ai-in-cyber-security-friend-or-foe-activity-7164257971500253184-Xh2o
- https://www.planetcompliance.com/ai-and-ml-in-cyber-security-management
- https://www.networkbridge.co.uk/blog/the-role-of-artificial-intelligence-in-cybersecurity
- https://www.mltut.com/random-forest-algorithm-a-complete-and-super-easy-guide
- https://zawyatech.com/applications-of-machine-learning-in-cybersecurity
- https://www.opastpublishers.com/open-access-articles/revolutionizing-industries-the-impact-of-artificial-intelligence-technologies-5805.html

- https://www.amitjadhav.com/blogs/AI-From-Basics-to-Current-Developments
- https://www.eschbach.com/en/resources/blog/2023-03-23_staying-safe-in-the-cloud.php
- https://slashdot.org/software/intrusion-detection-and-prevention
- https://techpreptalks.com/forrester-top-5-cybersecurity-threats-for-2024-are-you-prepared
- https://www.algomox.com/usecases/security-and-threat-management/zero-day-a◎ack-recognition
- h◎ps://cyberpedia.reasonlabs.com/EN/trojan%20detection.html
- h◎ps://identitythe◎.org/website-security/firewall
- h◎ps://www.purevpn.com/blog/what-is-a-botnet
- https://medium.com/@21varsha.shri.m/neural-networks-bdb7d76e3e1a
- https://agileblue.com/unlock-the-potential-of-ai-in-safeguarding-your-cybersecurity
- https://votiro.com/blog/glba-cybersecurity-and-your-business-conquering-hidden-threats-with-cdr
- h◎ps://www.anodot.com/blog/what-is-anomaly-detection
- https://www.knowledgehut.com/blog/data-science/machine-learning-for-anomaly-detection
- https://dipranjan.github.io/dsinterviewqns/contents/Algorithms/Anomaly%20Detection.html
- https://medium.com/@raghavsriram/uncovering-unusual-customer-behaviors-anomaly-detection-with-clustering-techniques-15895c04dcfc
- h◎ps://blog.uniqkey.eu/detect-and-prevent-insider-threats
- https://www.linkedin.com/pulse/how-can-ai-prevent-insider-threats-samuel-ajiboyede
- h◎ps://moldstud.com/articles/p-exploring-the-relationship-between-artificial-intelligence-and-cybersecurity

- https://fronty.com/Artificial%20Intelligence%20&%20Cybersecurity:%20Boosting%20Digital%20Defense
- https://rootstack.com/en/blog/artificial-intelligence-and-data-cybersecurity-fighting-evolving-threats
- h◎ps://www.linkedin.com/pulse/how-ai-reshaping-way-we-develop-optimize-automate-so◎ware-dhawan-i21uc
- https://www.harrisonclarke.com/blog/mlsecops-revolutionizing-cybersecurity-incident-response
- https://www.coursesidekick.com/information-systems/4756902
- https://theintactone.com/2023/04/08/concept-of-computer-forensics
- h◎ps://www.linkedin.com/pulse/digital-forensics-sajin-pd-26otc
- https://www.linkedin.com/pulse/what-digital-forensics-vulsanx-vqiec
- McAfee, A. (2024). The arms race in AI cybersecurity: How artificial intelligence is changing the threat landscape [8]. John Wiley & Sons.
- Wang, W., Ji, Q., Liu, Y., & Zou, Z. (2023). Adversarial machine learning in cybersecurity: A survey [9]. IEEE Transactions on Computational Intelligence and AI in Games, 1-17.
- https://agiaccelerate.com/blog/ai-enhanced%20cybersecurity%20protecting%20digital%20assets%20and%20infrastructure%20in%20a%20connected%20world
- https://www.linkedin.com/posts/lakmini-rathnayake-10974b230_quantum-computing-are-we-on-the-verge-of-activity-7110094835222552576-E8nX
- Slaughter, S., Wa◎s, W., & Bridgewater, J. (2020). Managing the risks of artificial intelligence [5]. International Security, 45(1), 32-79.
- Shao, J., Zheng, Y., Tang, X., & Zhao, Y. (2023). Deepfake detection: A survey [6]. IEEE Transactions on Circuits and Systems for Video Technology, 33(1), 321-340.

- Li, Y., Li, Z., & Tian, Y. (2022). A comprehensive survey on AI-powered phishing a◎acks. ACM Computing Surveys (CSUR), 55(2), 1-34.
- Minerva, R., Lightfoot, D., & Li, X. (2023). Artificial intelligence of things: A survey on the convergence of AI and IoT. ACM Computing Surveys (CSUR), 56(2), 1-38.
- Singh, J., Patel, V., & Jin, Y. (2023). Artificial intelligence for next-generation data encryption: A comprehensive survey [2]. IET Networks, 12(2), 123-130.
- Lee, I., & Kim, J. (2023). The challenges of securing the Internet of Things (IoT) and artificial intelligence of Things (AIoT): A review. Sensors, 23(7), 8425.
- Wachter, S., Mi◎elstadt, B., & Floridi, L. (2020). Transparent AI: An inquiry into its possibilities and limitations. Cambridge University Press.
- https://www.terranovasecurity.com/blog/ai-in-cyber-security
- https://dai-global-digital.com/unlocking-the-potential-of-generative-ai-in-cybersecurity-a-roadmap-to-opportunities-and-challenges.html
- https://dai-global-digital.com/unlocking-the-potential-of-generative-ai-in-cybersecurity-a-roadmap-to-opportunities-and-challenges.html
- https://www.linkedin.com/pulse/present-future-artificial-intelligence-cybersecurity-cybalt/
- https://www.kaspersky.com/enterprise-security/wiki-section/products/machine-learning-in-cybersecurity
- h◎ps://www.aiio◎alk.com/machine-learning-examples-in-real-life/
- https://www.researchgate.net/figure/Encryption-and-Decryption-in-Cryptography_fig1_350123908
- https://swimlane.com/blog/the-role-of-preparation-and-process-in-incident-response/
- h◎ps://edafio.com/blog/the-8-most-common-cybersecurity-a◎acks-to-be-aware-of/

- h◎ps://www.linkedin.com/pulse/next-generation-firewall-ngfw-one-ossama-rashad/
- https://www.liquidweb.com/blog/host-based-intrusion-detection-system/
- https://www.researchgate.net/figure/System-architecture-for-feature-extraction-and-pattern-recognition-used-by_fig1_341208194
- h◎ps://www.logpoint.com/en/blog/insider-threat/
- h◎ps://www.cyber.airbus.com/automating-the-soc-towards-ai-based-incident-response-in-the-factory-of-the-future-2/
- https://www.servicenow.com/products/legal-service-delivery/what-is-digital-forensics.html
- https://www.linkedin.com/pulse/how-implement-vulnerability-management-process-your-ebenezer/
- https://www.frontiersin.org/articles/10.3389/fcomp.2022.1070493/full
- https://medium.com/@Shahriare/ai-bias-and-fairness-58c277bc511f
- h◎ps://openeducationalberta.ca/educationaltechnologyethics/chapter/ethical-considerations-when-using-artificial-intelligence-based-assistive-technologies-in-education/
- https://www.veritis.com/blog/ai-and-iot-collaboration-in-addressing-industry-challenges/
- https://www.veritis.com/blog/ai-and-iot-collaboration-in-addressing-industry-challenges/
- h◎ps://www.agari.com/wp-content/uploads/2021/05/cyber-threat-intel-blog-lifecycle.png
- h◎ps://www.linkedin.com/pulse/soarsecurity-orchestration-automation-response-nikhil-gadgil/
- https://www.researchgate.net/figure/Conceptual-map-of-security-orchestration-and-automation_fig1_344260727

Made in United States
Orlando, FL
04 November 2026

90201576R00140